# Degradation of Natural Building Stone

Proceedings of 2 sessions sponsored by the
Rock Mechanics Committee of
The Geo-Institute of the
American Society of Civil Engineers
in conjunction with the ASCE Convention
in Minneapolis, Minnesota

October 5-8, 1997

Edited by J. F. Labuz

Geotechnical Special Publication No. 72

Published by the
**ASCE** *American Society of Civil Engineers*
1801 Alexander Bell Drive
Reston, VA 20191-4400

Abstract:
This proceedings, *Degradation of Natural Building Stone*, represents an effort to collate the various disciplines working in the area of building stone degradation. Structures composed of natural building stones are subjected through time to mechanical loads, temperature variations, and chemical attacks that result in degradation of the material in the form of microcracks. Damage diagnosis of natural stone is the first stage of planning needed for proper remedial action and restoration. To this end, case histories and successful experimental procedures are emphasized in this proceedings. It is the hope that this publication will encourage a dialog among civil engineers, contractors, and architects in order to develop efficient, economic, and safe restoration and preservation policies.

Library of Congress Cataloging-in-Publication Data

Degradation of natural building stone : proceedings of 2 sessions by the Rock Mechanics Committee of the Geo-Institute of the American Society of Civil Engineers in conjunction with the ASCE Convention in Minneapolis, Minnesota, October 5-8. 1997 / edited by J.F. Labuz.
    p.     cm. -- (Geotechnical special publication; no. 72)
Includes index.
ISBN 0-7844-0279-5
1. Building stones--Testing. 2. Building stones--Deterioration. I. Labuz, J. F. II. American Society of Civil Engineering. Rock Mechanics Committee. III. ASCE National Convention (1997 : Minneapolis, Minn.) IV. Series.
TA427.D44 1997          97-36205
620.1'322--dc21  CIP

# GEOTECHNICAL SPECIAL PUBLICATIONS

# FOREWORD

The aesthetics and longevity of natural stones are without question. Nonetheless, buildings and historical monuments composed of natural stone are subjected through time to mechanical loads, temperature variations, and chemical attack, which may lead to degradation of the material. Degradation is manifested by damage in the form of microcracks. The cover photograph shows these tiny cracks within an otherwise solid rock. The cracks are very small, but their number is such that the elastic modulus can be reduced by 50% (that is, if the cracks could be removed the modulus would be twice as large). However, the presence of microcracks does not mean that the material is incapable of exhibiting high strength and stiffness; for example, the rock shown on the cover has mechanical properties similar to some metals. Rather, a localization of microcracks due to various loading conditions leads to degradation, and it is this increase in damage that must be quantified.

Damage diagnosis of natural stone is the first stage of planning remedial steps to which the success of the restoration will be entrusted. An erroneous diagnosis may be very harmful both to the structural and the economic outcome of the operation. The restoration processes applied throughout the world to facades of historical structures and monuments of cultural heritage involve, up to now, site specific strategies, based on evaluation procedures for mechanical properties well known to the civil engineer, but of little value to the contractor and poorly understood by the architect. It is imperative then, that discussion on quality assessment of natural stone proceed on a level where the various parties can interact, a level found at an ASCE Convention.

Two full sessions on Degradation of Natural Building Stone were held at the 1997 ASCE Annual Convention and Exposition (October 5-8, 1997, Minneapolis, MN). An emphasis was placed on case histories and on successful experimental procedures. The objective of the sessions was to establish a dialog that will eventually lead to efficient, economic, and safe restoration and preservation policies, as well as procedures related to new construction. This objective paralleled with the theme of the Convention — Innovative Civil Engineering for a Sustainable Development.

This volume represents an effort by the Rock Mechanics Committee of the GeoInstitute of ASCE to collate the various disciplines working in the area of stone degradation. Quarry operators, government and university researchers, and private consultants have contributed to this volume. The papers were reviewed by anonymous referees according to guidelines established by the ASCE Journal of Geotechnical and Geoenvironmental Engineering. As a result, the papers are eligible for ASCE awards, as well as discussion within the Journal of Geotechnical and Geoenvironmental Engineering. The help of Gregg A. Scott (chair of the Rock Mechanics Committee) and Priscilla P. Nelson (contact member from the Board of Governors of the Geo-Institute) is gratefully acknowledged.

**J.F. Labuz**
**Department of Civil Engineering**
**University of Minnesota**
**July 1997**

# CONTENTS

# MODERN DIMENSION STONE QUARRYING
# AND FABRICATION TECHNOLOGIES

## C. J. Muehlbauer[1]
## J. P. Fuchs[2]

## Abstract

The harvesting of natural stones for use in building construction is one of man's oldest crafts. The properties and behavior of natural stone materials have not changed since man's early endeavors in manufacturing usable products from these raw materials. While the raw product is unchanged, the means and methods of harvest can and have been dramatically improved. These improvements have resulted in commercial, safety, quality, and environmental benefits.

## 1.0 Introduction

It is not possible within the limitations of this paper, to provide comprehensive discussion of every aspect of the stone manufacturing process. It is the intent of the authors to provide a cursory overview of the major steps involved in procuring and refining natural stone products. Since both authors are employed by the same company, the means and methods described in this paper are essentially those in current use by that company. As in any industry, the equipment and procedures used will vary between producers.

## 2.0 Selection of the Quarry Site

Most quarries start out as a natural outcropping of stone exposed by an uplift or erosion. When the prospector initiates evaluation of the stone deposit for commercial use, a myriad of questions are waiting to be answered before the site can be deemed a viable commercial investment.

[1]Industrial Products Division, Cold Spring Granite Company, 202 South Third Avenue, Cold Spring, Minnesota, 56320 USA
[2]Quarry Engineer, Cold Spring Granite Company, 202 South Third Avenue, Cold Spring, Minnesota 56320 USA

1

The most critical concern is the marketability of the material. Samples will be taken from the site and distributed to industry professionals for input regarding the potential market demand for the material in a variety of products. A material with marketability in multiple product lines has a distinct commercial advantage, as the cyclical natures of different markets tend to balance each other and provide a more constant demand for the product. This will enable the quarrier to maintain uninterrupted operations on the site. If similar colored materials already exist in the marketplace, the price available for the new material will be governed by the current prices of the existing materials.

Of nearly equal concern are the soundness and consistency of the rock deposit. Quarries with many natural cracks and fissures will limit the obtainable block size and reduce the production yield. Both small block sizes and reduced yields will result in high costs which may not be recoverable in sales. Rapid changes in color or grain size will limit the end use of the material to products that do not require color matching from stone to stone. Overburden, or the "non-harvestable" earth and rock that cover the deposit will have a significant impact on the cost of extraction and future reclamation. Test cores will be drilled throughout the deposit to determine soundness, consistency, and overburden.

The potential quarry site may lie within an environmentally sensitive area. Research must be completed on what regulatory agencies govern the activities at the site and what environmental studies and reclamation efforts are required. Such mandates vary widely from region to region and dramatically affect operating costs.

If the proposed quarry site is in a heavily populated area, the appearance of a new quarry operation may be met with opposition. By contrast, in other areas, the new operation may be welcomed as a source of employment.

Most new quarries are started in sparsely populated areas, in fact, most new quarry sites are in extremely remote areas. Being separated from populated and industrialized areas initiates another set of concerns. The primary concern is the availability, cost, and quality of the local labor. Additionally, the availability of parts and service for the heavy equipment must be researched, as the quarrier cannot afford to experience long term work stoppages due to the equipment problems.

### 3.0     Traditional Quarry Methods

There has always been a requirement for some form of mechanical advantage for man to move the stone within and around the quarry site. For centuries, this mechanical advantage was provided by the "derrick", which was the nucleus of nearly every major quarry operation.

The two types of derricks commonly found in the stone industry were the "stiff-leg derrick", and the "guy cable derrick". Both types of derricks were originally constructed of large straight logs which were later replaced by steel. Both types of derricks featured a member with a pivot point at the base and a block and tackle attached to the other end, making the boom section of the derrick. The mast section stood vertically, with a pivot point at the base concentric with the pivot point of the boom section, and a block and tackle on the top which was used to raise and lower the boom. The difference in the two types of derricks was the method used to support the mast and boom.

The stiff leg used two braces, both attached at the top of the mast with the other ends secured to a concrete pier or granite ledge. These two braces were usually set about 100 degrees apart. The guy cable derrick used a series of cables (or ropes prior to the availability of cable) with one end attached to the top of the mast and the other end secured to the granite ledge below. The guy cables were positioned in 40 to 60 degree increments completing a full 360-degree circumference around the base.

The most recently engineered derricks boasted tremendous lifting capacity, with some of the larger derricks in the last 40 or 50 years having lifting capacities of 150 to 200 tons with booms extending in excess of 60 meters from the base. Large quarry operations would have multiple derricks erected with overlapping radii of coverage. Derricks would also be erected at multiple levels making hand-offs from one level to the next to lift stones the full height of the quarry walls.

The derrick system of quarrying was a very labor intensive method of moving material, equipment, and in some cases, quarry workers. Precious time was wasted in the handling of the stone, and safety was always a concern as blocks of stone were lifted more than 30 meters off of the quarry floor. The derricks were essentially immobile, and quarry size was limited to their reach. The industry was searching for a safer and more cost effective way to mechanize the quarrying operation.

## 3.1   Drive-In Quarry Systems

Many of the best ideas developed in an industry appear so ridiculously simple that one cannot understand why it hadn't been thought of previously. The concept of the "Drive-In" quarrying system is such an idea.

Developed in the early 1970's in Finland, the drive-in system of quarrying is centered around one modern machine: a rubber-tired hydraulic front end loader. As the iron ore and coal mines saw the big cable operated shovels give

**Photo 1 - Front End Loader Transports a Block from the Quarry**

way to the hydraulic front end loaders, the dimensional stone quarries saw the large cable operated derricks give way to those same big front end loaders.

Modern front end loaders have capacities to lift and transport 50 ton blocks. While derricks limited the early quarries to approximately 100 meters square, the drive-in quarry is without lateral dimension limits. Some drive-in quarries measure in excess of 1 km square.

The drive-in quarry consists of a series of multi-level terraces, called "benches". Access to all of the benches is done via ramps cut from the natural stone beds. The slope of the ramps is quite steep, usually between 40 and 45%. The bench height can vary anywhere from 5 to 8 meters, with the optimum being about 5½ meters.

The drive-in quarrying system brought changes that would increase the rate of production of a given quarry by a factor of 4 or 5, which is often paralleled by a reduction in labor hours. Obviously, dramatic cost reductions were realized with the conversion to drive-in methods. As an example, a particular derrick quarrying operation

**Figure 1 - Typical Arrangement of Benches in a Drive-In Quarry Operation (Tamrock, 1995)**

in 1984 had 14 workers and produced approximately 600 m³ annually. A new site was prepared a short distance from the derrick quarry in 1986 and developed as a drive-in system. Today the new quarry produces approximately 2,800 m³ with 8 workers. The production per man-hour is now 8 times that of the original derrick quarry! All of the commercial advantages have been paralleled with increased safety, as the blocks of stone are lifted a maximum of a few meters above the surface.

## 3.2    Separation of the Stone Masses

The granite in a quarry is normally removed in large sections, called "loaves". The size of the loaf will vary depending on the material type and quarry dimensions. The vertical dimension is governed by the bench height, which is usually between 5 and 8 meters. The depth, or distance back from the exposed face of the loaf is usually about 6 meters. The length of the loaf varies widely from quarry to quarry, and can be anywhere from 10 to 50 meters.

The first step in separation of the granite loaf is to free the ends of the loaf. This is accomplished by "burning", diamond abrasive cable saws, or high-pressure water cutting:

The most frequently used method is to burn the stone away with a cutting tip using fuel oil and compressed air to support an intense flame. The flame doesn't actually burn the granite, but simply takes advantage of the stone's inability to accommodate differential heating. The exposed crystals expand rapidly when heated by the flame and break away from the stone mass in flakes. The flame cuts a trench about 100 mm in width. Because of the localized heating, it has the potential to drive cracks into the adjacent stone which will reduce yields. Additionally, it is not effective in all types of granite. There must be a sufficient amount of quartz present in the granite for the flame to produce the necessary thermal desegregation in the rock.

The second option to complete this task, is the use of a diamond abrasive cable or belt. A vertical hole is drilled from the top of the bench and a horizontal hole is drilled into the face of the bench to intersect with it. A diamond cable is strung through the holes to form a loop and the sawing of the kerf begins. Diamond abrasives have been used in finish fabrication operations for decades, but have just recently, about the last 15 years, been available for quarrying operations. In softer materials such as limestone, sandstone, some gneiss's, and anorthosites the diamond works extremely well. As materials approach about 7.0 on the Mohs scale of hardness, the diamond segments become less effective and service life of the segments is reduced. Another concern with diamond segmented cables is that there will frequently be some unresolved stresses in the natural mass of stone. When portions of the stone mass are removed, the stresses seek to neutralize themselves and in doing so produce cracks, commonly referred to as "pressure cracks". If one of these cracks develops during sawing with a diamond cable, the granite mass can shift and pinch the cable unrecoverably in its own kerf. Since the average cost of a diamond segmented cable is about $8,000.00 USD, this does create some concern.

The newest technology available for this task is the high pressure water jet cutter. Like diamonds, it is an existing technology, but hasn't been applied to quarrying efforts until recently. The machine consists of a diesel engine to run a hydraulic intensifier which in turn creates water pressures of 275 MPa (about 40,000 psi). This method of cutting will penetrate nearly any material it encounters. As an example, a flame cutter would have great difficulty cutting through a basalt dike, but the water jet cutter will easily abrade the material away. Water cutting does not induce any thermal expansion or cracking into the rock, and it is not hampered by pressure cracking of the quarry. The resultant cut is quite straight and accurate. Advantages are offset by financial concerns, as both the initial investment and the operating cost of this machinery are very high. Additionally, the method is not usable during the winter months in the northern regions of the country.

Once the two ends of the loaf are cut free, the balance of the stone separation is done with explosives. The two planes that remain are separated in one effort, commonly called "L-Blasting". To facilitate the separation, the stone is perforated with holes which both define the separation plane and accommodate the loading of the explosive. The drilling of the holes is accomplished with a percussive carbide tipped bit under hydraulic force. The stone dust is then collected in an integral dust collection unit on the drilling machine. Modern drills can drill at feed rates of about 1.25 meters of travel per minute.

Horizontal holes are drilled about 10 cm above the quarry floor into the face of the loaf. On top of the loaf, vertical holes are drilled about 6 meters back from the face of the loaf downward. The horizontal and vertical holes will come to within a few centimeters of intersection. The holes are usually 32 mm in diameter, and are spaced on about 200 mm centers (this distance varies due to the ease and regularity with which the granite will split). The horizontal and vertical alignment of the holes is critical, as this will control how evenly the loaf will separate from the mother mass of granite, as well as the squareness and regularity of the finished blocks.

Historically, black powder was used as an explosive in our industry. Black powder is indeed an explosive, but it was developed and formulated as a propellant. In its traditional application, black powder users wanted a lower peak pressure with a great volume change. The dimensional stone quarrier wants just the opposite: a rapid pressure increase with minimal volume increase. Dynamite has been tried but cannot be used with confidence due to its tendency to drive cracks into the stone. There are several varieties of "low grade dynamite" that have proven to be the most reliable. The commonly used trade names are "K-Pipe" (manufactured in Finland), and the lower cost domestic products "Stonecutter", and "Dynashear". All of these materials are manufactured in stick form.

The amount of explosive required to free a loaf is calculated based on the mass of the loaf that is to be freed. Charges are loaded in the horizontal holes by tying a detonation cord to a charge and pushing it to the full depth of the hole. Additional charges are pushed into the holes as required to achieve the

Photo 2 - A freshly separated Loaf of Granite (Rockville, MN)

correct amount of explosive in a uniform pattern. The vertical holes are loaded from the top with the charges tied to the detonation cord to maintain vertical spacing. Once the charges are loaded, the detonation cords are linked by a "trunk line" detonation cord. The trunk lines for both the vertical and horizontal hole charges are fitted with a blasting cap on the same end of the loaf. Ignition of the blasting caps is done electrically. The detonation cord is a very fast burning fuse with a burn rate of approximately 6,700 m/s. When blasting a 50 meter long loaf, the ignition of all individual explosive charges will occur within about 7 milliseconds. The relatively light charges force a crack from drill hole to adjacent drill hole in a successive pattern throughout the loaf. By design, the explosives are to develop enough pressure to displace the loaf about 200 to 300 mm, and experienced quarry crews will consistently achieve this. After the explosives are detonated, a post blast inspection is performed to ensure that all charges were detonated and no live charges remain within the loaf. The entire process of cutting the ends, drilling the vertical and horizontal holes, loading the charges, and blasting the loaf free takes less than two weeks to complete.

One of the risks of separating granite masses by explosive forces is the occurrence of an explosive induced crack. This can be caused by both over or under loading the explosive charge, or more commonly, by the presence of water in the drilled hole. A typical explosive induced fracture will manifest itself in the finished slab as a series of small cracks starting at the drill holes and radiating into the field of the slab in a fan-shaped pattern.

## 3.3    Modular Sizing

Yield is the most significant factor affecting the profits of the quarrier and fabricator of natural stones. The usable blocks taken from a quarry constitute a small percentage of the total amount extracted. The usable portion, or "yield", is typically between 10 and 40%, but in extreme cases it can be less than 5% of the total extraction.

It is in the best interest of the quarrier to size the blocks at a dimension that will net just large enough slabs to yield the finished pieces or multiples of the finished pieces for the project. In North America, the most commonly found curtainwall module is 1,524 mm (5'-0"). Conscious of this fact, the dimensional stone quarrier will lay out a major portion of the production to net that size. Allowing for waste areas at the surface of the block, the quarrier will produce blocks of nominal dimensions of 1.7 x 1.7 x 3.2 m (approx. 5'-6" x 5'-6" x 10'-6"). Most granites have a density of about 2,700 kg/m$^3$, so this standard modular block having a volume of 9.25 m$^3$ weighs about 25,000 kg, and is about the maximum that can be hauled on public roads without special permits.

Many building projects require sizes that will produce unacceptable amounts of waste if the standard block size is used. The quarry foremen are furnished with "stock lists" for all projects and will identify these atypical sized pieces. Benches will be specifically laid out and drilled to produce optimum sized blocks for use in all areas of a project.

## 3.4    Material Grading

The last process to occur at the quarry site is the grading of the material. This will usually be done by a very experienced individual with keen eyesight for color. The blocks will be wetted to intensify the color, after which they will be inspected for color variations, background color tones, inclusions, lines, cracks, seams, or any other property that would be unacceptable in a given product line.

Those of us who work in the stone industry appreciate the inherent variations that occur in this natural material, yet the very unique feature about a stone that we find attractive may offend our customer. In nearly every natural granite type, the clearest blocks with the most consistent color and least amount of veining will be considered the highest grade and command the highest price.

Different product lines have different requirements. For example, a block destined for monument production must be of very clear stock to prevent distraction from the script that will be incised in its face, but it is not necessary that the color match from one block to the next. A building project will typically allow more veining and natural marking because it is viewed from a greater distance, yet it is imperative to have large quantities of blocks with uniform background color to complete the project.

## 4.0    Fabrication of Standard Granite Products

Natural granite is manufactured into wide variety of products.  Monumental and memorial applications are probably the most commonly recognized forms, followed by building claddings, pavements, and landscape features.  Granite products are used extensively in interior wall, floor, and countertop applications, both in commercial and residential construction.  In addition to memorial and construction uses, there are several common industrial applications.  Granite is used to make "surface plates" which are used as a precision measuring datum or for precision machine mounts and inertia plates.  Industrial applications where chemical resistance is required often use granite products, as for the lining of sulfuric or hydrochloric acid tanks used in the "pickling" of steel, or for the construction of bromine extraction towers.  The discussion of fabrication in this paper is limited to building application products, which is the most standardized of the different product lines.

### 4.1    Slabbing

After the raw block has been inspected, graded, and transported to the fabrication site, the first process in fabrication is the "slabbing" of the block.

While there are generally standard slab thicknesses in the industry, available thicknesses and tolerances may vary slightly from one producer to another. Typically, 10, 20, 30, 40, 50, 80, 100, 150, and 200 mm nominal slab thicknesses are produced by most fabricators. Of these thicknesses, 30 mm is the most commonly used thickness for

**Photo 3 - Gang Shot Saw**

curtainwall cladding.  When necessitated by long spans, high wind loading, or low flexural strengths, 40 mm, 50 mm, or even greater thicknesses may be required.

Slabs are cut from the blocks by three methods; a reciprocating blade using steel shot as an abrasive media, a twisted wire carrying a silicon carbide slurry as an abrasive media, or a 3.5 meter diameter circular saw fitted with diamond abrasive segments.

Reciprocating saws, commonly called "gang saws" or "gang shot saws", are the most popular method of sawing slabs worldwide. A gang of up to 140 blades is fixed at constant increments to achieve the desired slab thickness. A mixture of steel shot, hydrated lime, and water is fed to each blade. The steel shot serves as the abrasive, while the water and lime hold the shot in suspension. The entire gang of blades, which weighs in excess of 30 tons with its fly wheel and eccentric arm, reciprocates together as one unit. Most of the modern gang saws oscillate in a pendulum motion instead of a straight line back-and-forth action. Energy usage is dramatically reduced by this design, as the reversal of momentum is provided by the pendulum motion and the energy consumption is limited to that which is required to overcome friction. The reciprocating saws provide astounding accuracy. Slab thickness can stay within tolerances of ±1.5 or even ±1.0 mm across the 2.5 to 3.0 m length of the slab.

| Mountain Green: | 40 mm/hr |
|---|---|
| Rockville: | 35 mm/hr |
| Carnelian: | 35 mm/hr |
| Charcoal Black: | 30 mm/hr |
| Wausau Red: | 25 mm/hr |

**Table 1 - Saw Feed Rates for Gang Shot Sawing of Granites**

One disadvantage of the reciprocating gang saw is the long time required to complete the sawing of a block. As different stones vary in abrasion resistance, the feed rate of the saw must be adjusted accordingly. Listed in Table 1 are some examples of different granite types and their corresponding saw feed rate. With this data, it is readily apparent that sawing through a granite block is a time-consuming process. Normally the complete sawing of one block, called a "saw-out", including loading and material handling, takes about 3½ days at 24 hours per day. For this reason, reciprocating saws are most efficient when sawing thinner slabs. The reduced surface area of the thicker slabs is not sufficient to overcome the costs of operating the machine for several days.

Wiresaws consist of a two-strand carbon steel twisted wire carrying a

**Photo 4 - Gang Wire Saw**

silicon carbide slurry as the abrasive agent. The wire is under a constant tension of 1.8 kN and moves at a speed of approximately 25 m/s. The direction of the twist in the wire is reversed every 30 m to prevent rotation and lateral drift. Wiresaws historically produced crude accuracies, often resulting in slab thickness variations of ±10 mm. In recent years they have been improved with better alignment and drift control. Modern wire saws can produce accuracies of ±3 mm.

The wiresaw's primary advantage is speed. A wire saw can cut through a granite block with a down feed rate of 300 mm per hour, enabling a complete sawing of a block in one eight-hour shift, including loading the block and removal of the sawn slabs.

The 3.5 meter diameter circular saw is usually limited to nonstandard thicknesses and specialty applications of low volumes. These saws produce reasonable accuracies (±3 mm) and can cut through a block with impressive speed (between 300 and 800 mm/hr), but do so at the highest cost of any of the slabbing methods. The 3.5 m saw blade is fitted with diamond segments which are cooled and lubricated with water. The blades turn at approximately 130 rpm, producing a rim speed of 24 m/s. The saw is lowered between 4 and 10 mm per pass, depending on the hardness of the granite. The combination of the use of diamonds, which are the most expensive abrasives used for stone cutting, and the fact that the 3.5 m saws produce only one slab at a time contribute to the high operating cost of this machine.

Photo 5 - 3.5 meter Saw

## 4.2 Face Finish

There are a wide variety of standard and proprietary face finishes available for natural granites. The most popular finishes for building granite panels are the thermal, honed, and polished finishes.

A thermal finish (also called flamed), is a rough textured finish produced by a rapid heat application. A propane flame is passed over the face of the granite slab causing the outer crystals to flake off, leaving a textured finish, sometimes with a sparkle effect. The degree of coarseness varies due to the size of the grains in the stone. In North America where we enjoy relatively low propane costs, the thermal finish is the most economical to produce, and is widely used in pavement applications due to its frictional properties as well as its economy. The ADA (Americans with Disabilities Act - Bulletin #4, 1990) suggests a static coefficient of friction of 0.60 or greater for level walking surfaces. Most thermal finish granites will demonstrate a static coefficient of friction (ASTM C 1028-89) of 0.70 or greater, comfortably complying with the recommendation.

**Photo 6 - A Thermal Finish is applied to a Granite Slab**

A honed finish is a smooth finish, free of visible scratches, but not visibly glossy or reflective. A honed finish is produced on an automated line with a series of rotating heads holding abrasive bricks which are held against the face of the stone with hydraulic pressure. The abrasive varies from course to fine as the slab progresses through the machine. Honed finishes are used where refined appearance is desired, but without the distractions of the reflections occurring in polished surfaces.

A polished finish is a high gloss, highly reflective surface. By polishing the granite, the different minerals comprising the stone will demonstrate distinctive colors and many will be identifiable with the naked eye. This finish is produced with the same machinery as the honed surface, but with two additional heads. These heads use an aluminum oxide abrasive compound to produce the final polish of the granite.

**Photo 7 - A Polish Finish is applied to Granite Slabs**

**Photo 8 - Continuous Cutting Saw**

### 4.3    Panel Sizing

Once the slabs are cut to thickness and a finish has been applied to one surface, the next step in the fabrication sequence is to cut the slabs to specified dimensions.

The slabs undergo another grading process, where subtle color ranges are sorted so that the color variation can be minimized for a particular project. In some materials, color can vary so rapidly that opposite ends of the same slab may be destined to two separate projects. The slabs are then manually marked with a rough "layout" of what finished panels are to cut from them. All panels on a project are assigned a unique number. It is at this point when the unique identifying number is assigned to the particular piece of stone.

The slabs will be cut to a final face dimension by saws using circular blades with diamond abrasive segments. The diamond segments are cooled and lubricated with water. Saws with a singular blade will be guided manually by the sawyer aided by a laser sighted alignment device. When the panel sizes are repetitive, the cutting will be done with a multiple bladed "continuous cutting saw". These saws typically can be programmed by the operator after which the blade positioning is monitored electronically. Thin slabs of 20 or 30 mm can be sawn through with a single pass of the diamond blade. Thicker slabs require multiple passes in the same kerf to cut through the entire depth of stone.

### 4.4    Edge Treatments

Atypical conditions require specific edge treatments to be made to the stone panels. Most commonly, these edge treatments require either applying a finish to the edge or a miter cut to accommodate an exterior corner.

Due to stones susceptibility to breakage, miter cuts are performed in natural stones differently than in other materials. The miter joint in stone has a blunt nose instead of a point to reduce the potential of chipping. This detail is called a "quirk miter" (Fig.2).

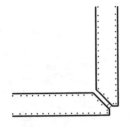

**Figure 2 - Quirk Miter Corner Detail**

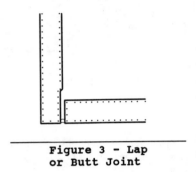

**Figure 3 - Lap or Butt Joint**

When a "lap joint" or "butt joint" (Fig. 3) is used at a corner condition, one panel requires a finish on the edge to match the face finish. Depending on the sawing method and equipment used, the stone slabs can vary by as much as ±3 mm. It would be very unsightly to position two slabs at opposite extremes of this tolerance adjacently with exposed edges. To avoid this, the edges must be calibrated to a uniform thickness. In the case of a honed or polished finish, the edge is finished and the thickness calibrated in one pass of a machine. In the case of a thermal panel, the finish is applied by a hand-held torch after which the thickness is calibrated in a separate operation.

### 4.5  Anchorage Preparations

There are many varieties of anchors used in natural stones. The most commonly used anchors for building panels are one of three types:  Local edge anchors;  Continuous edge anchors;  or Back anchors.

Local edge anchors are usually made either of stainless steel straps with a bent end penetrating the stone or a combination of a stainless steel rod and plate which penetrates the stone. These anchors resist lateral loads only while the gravity loads are addressed by other means. Only a short length of kerf cut in the stone's edge is required to facilitate engagement. These short kerfs are plunge cut with a small (150 mm diameter) circular blade using diamond abrasive segments.

Continuous edge anchors usually consist of an aluminum extrusion. These anchors are usually designed to resist both gravity and lateral loads. The aluminum section requires two continuous kerfs cut in opposite edges of the stone. The kerfs are cut by a semi-automated machine which holds the stone in a fixed position while a diamond segmented blade cuts the kerf in a single pass.

Back anchors, as their name suggests, are normally located in the back surface of the stone, and are designed to resist both gravity and lateral loads. A common back anchor is the Cold Spring Granite Company Type Number 31 Anchor. The anchor consists of a stainless steel bolt fitted into a dovetail shaped slot which is routed in the stone. The most significant advantage of this anchor type is quality control. Being "dry" assemblies, back anchors are not dependent upon any cementitious or polymer fillers, which lessens the probability of a field error during installation. A less obvious advantage is the ability to locate the anchor at optimum placements in the panel. Perimeter anchorages can result in high flexural stresses in

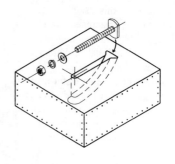

**Figure 4 - Type No. 31 Anchor**

a stone panel by requiring it to span its full face dimension. By proper placement of back anchors, the designer can take advantage of the contra flexure of the panel to resist wind loads. This will allow the use of either larger panels or lesser thicknesses without exceeding flexural stress limitations.

## 4.6    Contoured Faces

The majority of building products consist of flat, rectangular panels, and most of the development in automated machinery has been focused on the repetitive tasks involved in producing them. Interesting challenges await the stone fabricator when the design professional deviates from the rectilinear patterns and introduces curved elements into the design concept. It is quite frequent that a building design incorporates a circular or arced feature. Granite panels with arced faces are fabricated with two different pieces of machinery.

The traditional method involves a wiresaw using essentially the same wire and silicon carbide abrasive as the wiresaws used for slabbing. The block is fixed in a cradle while a single strand of wire carves a path to produce the desired profile. A full-sized physical template is traced onto the end of the block. The operator controls the sawing unit manually with the ability to move it laterally as the wire travels downward.

The more modern method of producing this same piece involves a machine using a heavy cable (11 mm diameter) with cylindrical diamond abrasive segments. The diamonds cut much faster than the traditional wire and silicon carbide and leave a much smoother surface. The machine is programmable to produce a given curve, and will accept standard DXF files which can be transferred from CAD drawings. A profile detailed by a draftsperson can be simply loaded into the machine and reproduced in the stone.

**Photo 9 - Diamond Cable Saw**

Granite columns have always been a focal point of classical architecture. Traditionally, the columns were solid, load bearing units which took advantage of the material's greatest property: its resistance to compression. Modern construction schedules seldom allow for the delivery and erection of true, load bearing granite columns, so the visual effect is synthesized with a granite veneer surrounding a structural column of either steel or concrete.

In either case, the column is both shaped and finished on a lathe. A solid core column is held in place with fixtures penetrating alignment holes drilled in the ends of the stone. The pieces of a veneered column are clamped in a mandrel. The lathe turns at a variable speed to be adjusted for the diameter of the work piece as well as the finishing function being performed. Initial shaping is accomplished with pneumatic driven diamond grinding wheels held in place by temporary fixtures. Final finishing is usually done by hand-held tools as the work piece is spinning.

### 5.0    Computerized Product Tracking Systems

The advent of affordable and powerful computers has affected nearly every industry in current times, and the stone industry is no exception in this regard. Being a natural material, no two pieces of granite are identical, just like no two pieces of oak trim are identical. Within the stone industry, every piece of material produced is regarded as a unique product. In paint or textile production, attention is paid to the lot number of a product as two different lot numbers may vary in color. In the granite industry, every individual piece of material is essentially its own lot number. For this reason, two pieces of cladding with identical dimensional and detail specifications will be assigned unique identifying numbers to prevent exchange during installation.  Material is graded and color-blended at the fabrication site, and interchange of panels during installation can produce drastic color variations between adjacent panels.  Management of this amount of data is only achievable by computer.  The use of computerized production tracking systems have aided both scheduling and quality control.

The tracking starts at the quarry. Once a block of granite is trimmed to a roughly rectangular shape it is considered a product. It is tagged with a weather resistant label that identifies it as a unique product.  The numerical identification is traceable to the quarry from which it was produced, the date it was tagged, the rough dimensions, and approximate weight of the block. The quarry records can locate the original position of this block in the natural stone deposit.  The numbers of adjacent blocks can also be determined from the quarry records.

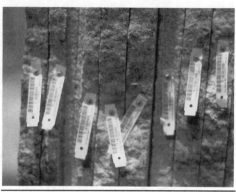

**Photo 10 - Slab Labels**

Once the block has been cut into slabs, each individual slab is tagged with a weather resistant label identifying it as a unique product.  This number is recorded electronically with the block from which it was sawn, the dimensions of the slab, and the date it was produced.

The slab will receive a face finish and will be cut to final size per the requirements of a particular project.  The finished piece of stone will be assigned a unique number, which is also recorded electronically with its dimensions, the slab number from which it was cut, and the date, work station, and responsible technician for all significant operations performed on that piece of granite.

Maintenance of a system such as this is a time consuming and costly process, but its contribution to quality assurance is immense.  If an inspector notes that an anchor hole was cut shallow, all pieces that require the same anchor that were cut on the same shift can be recalled via the computer's records and reinspected for proper anchor depths.

If a problem with the raw material is detected, the computer tracking system can be used to recall related pieces of stone.  Given the unique piece number of the product, the tracking system can identify the slab number from which it was cut and identify the other piece numbers that were taken from the same slab.  Reentering that slab number, the tracking system will identify the block number from which it was sawn and identify the sibling slabs taken from the same block. With those slab numbers, the numbers of all finished pieces cut from the entire block can be identified and recalled for inspection. The slab numbers cut from those blocks can be identified and the finished pieces cut from those slabs can be identified.

## 6.0    Safety in the Industry

Modern equipment has been designed with safety as a primary goal incorporating acoustic barriers and dust collection systems for operator protection . "Safe Zones" are maintained around all equipment to prevent accidental contact with moving parts. Both audible and visual warning systems are in place where workers and heavy equipment must share common space. Perimeter cables are required at quarry benches to prevent accidental falls. Mechanical assistance for lifting product and equipment is provided at all work stations. Employers provide personal safety equipment such as hard hats, safety glasses, respiratory devices, and hearing protection and hold regular safety training classes to stress the importance of proper and continual use of the equipment. Companies have instituted safety awareness programs with financial incentives to spark interest in safety amongst the employees.

**Photo 11 - Finished Panel is being handled with overhead hoist and vacuum pads.**

## 7.0  The Industry and the Environment

The overall awareness of the environment has been escalated to a higher plateau than ever before. Water filtration and reclamation facilities have been incorporated into fabrication facilities within the last 20 years. The solid particles (mostly silicon carbide and granite dust) are filtered out and sold as byproducts to other industries. The pH level of the filtered water is checked and acidic levels neutralized if required prior to return for use in the fabrication operations.

"Grout", or quarry waste, is being addressed upon the inception of a new quarry operation as opposed to addressing it only after the abandonment of the site. Reclamation plans for the site include the covering of the grout piles with topsoil and the planting of indigenous species of plants. The modern quarry plans must include closure details as well as operational details.

The waste fragments generated by the fabrication plant are being transported to reclamation sites where they are crushed for use as landscaping materials and road construction aggregates.

## 8.0 Conclusions

The many advancements made within this industry in the past several decades have provided the consumer with both lower cost and higher quality products. Continual research and development from global sources is expected to build upon the advancements already realized. Some of the new technologies are already in experimental use at this time. With these advancements, the use of the oldest building material known to man will assuredly be a popular choice for centuries to come.

## 9.0 References:

Americans with Disabilities Act - Bulletin # 4 (1990), U.S. Architectural and Transportation Compliance Board.

ASTM C1028-89 Standard Test Method for Determining the Static Coefficient of Friction of Ceramic Tile and Other Like Surfaces by the Horizontal Dynamometer Pull-Meter Method

Tamrock Oy, FIN-33311 Tampere, Finland (1995) Quality Drilling Controlled Quarrying

# STONE EXPOSURE TEST WALL AT NIST[1]

Paul E. Stutzman and James R. Clifton
Building and Fire Research Laboratory
National Institute of Standards and Technology

## ABSTRACT

The National Institute of Standards and Technology (NIST) Stone Test Wall was originally built in 1948 at the old National Bureau of Standards (now NIST) site in Washington D.C. It was moved intact in May 1977 to its present site at NIST in Gaithersburg, MD. The purpose of the stone test wall is to study the performance of stone subjected to weathering. It contains 2,352 individual samples of stone, 2032 of which are domestic stone from 47 states, and 320 stones from foreign countries. Over 30 distinct types of stones are represented, some of which are not commonly used for building purposes. There are many varieties of the common types used in building, such as marble, limestone, sandstone, and granite. Unexposed specimens have been stored indoors at room temperature and humidity facilitating the determination of weathering effects.

The type and extent of degradation observed in the stone specimens varies greatly, ranging from deep erosion to mild scaling (sugaring). In general, specimens of stones commonly found in buildings and monuments exhibited minor or slight degradation, usually some discoloration or scaling. The types and extent of degradation are described, with emphasis on common building stones. Also, ways to quantify the amount of degradation using computerized imaging techniques are discussed.

Keywords: building stone, degradation, discoloration, stone, stone wall.

---

[1] Contribution of the National Institute of Standards and Technology. Not subject to copyright in the United States.

## 1. INTRODUCTION

In 1880 the Census Office and the National Museum in Washington, D.C. conducted a study of building stones of the United States and collected a set of reference specimens from working quarries. This collection was merged with the Centennial Collection of U.S. building stones after the exhibition at which it was first displayed, the centennial exhibition in Philadelphia in 1876. A compilation of descriptions of producing quarries, commercial building stones, and their use in construction across the country was prepared for the report of the 10th census of the United States in 1880 [1]. This collection of stones, now augmented with building stones from other countries, was placed on display in the Smithsonian Institution.

In 1942 a committee was appointed to consider whether any worthwhile use could be made of the collection. It was decided that a study of actual weathering on such a great variety of stone would give valuable information. A plan was developed for building a test wall at the National Bureau of Standards (NBS) as a cooperative study between NBS and ASTM Committee C-18 on Building Stone. Subsequently, in 1948 a test wall was constructed at the NBS site in Washington D.C. (Figure 1). The move of NBS to Gaithersburg, MD in the middle 1960's and the occupancy of the old NBS site by the University of the District of Columbia, placed the wall in jeopardy, necessitating the move of the wall. It was moved intact in May 1977 to its present site at NBS (now the National Institute of Standards and Technology (NIST)) in Gaithersburg, MD (Figure 2).

The wall provides a rare opportunity to study the effects of weathering on different types of stones, with the climatic conditions being the same for all stones. It offers a comparative study of the durability of many common building stones that have been used in monuments, commercial, and government building. Also, the wall has served to preserve a valuable collection of building stone and should be useful as a reference for builders in identifying the kinds of stones which may be locally available. As the wall is approaching 50 years of age, interesting degradation features are being observed. Important features of the stone wall, its design and construction, and the exposure conditions are first described in this paper, followed by a description of the extent and types of stone degradation, with emphasis on common building stones. Ways to quantify the amount of degradation using computerized imaging techniques are also briefly discussed.

Figure 1. The stone test wall a few months after construction in 1948. The limestone and dolostone coping appears to have darkened soon after construction.

Figure 2. The stone test wall at the NIST Gaithersburg location (1997).

## 2. FEATURES OF THE STONE WALL

### 2.1 Construction Features

The wall is approximately 11.6 m (37 ft 9 in) long, 3.9 m (12 ft 10 in) high and
0.61 m (2 ft) thick at the bottom and 0.3 m (1 ft) at the top. The south face
includes a granite base, limestone coping, marble quoins on the west end, and
sandstone quoins on the east end. In its original location the wall was built on a
concrete foundation which extended 0.76 m (30 in) below grade. The limestone
coping prevents precipitation from infiltrating the wall from above.
Waterproofing against migration of runoff and underground water is provided by a
barrier of Virginia slate, portland cement mortar, and copper sheeting. Similar
waterproofing features were incorporated into the wall when it was placed at its
new site. It is now situated on the top of a mound which facilitates the drainage of
runoff precipitation.

As shown in Figure 2, the front of the wall was divided into separate sections, with
a divider strip extending 200 mm (8 in) from the front side into the wall. Two
kinds of stone-setting mortar were used in the front of the wall. The stones above
the base in the East side were set in a 1:3 lime mortar, using a high calcium
hydrate. Stones in the west half , including stonework on the back, were set in
1:0.4:3 portland cement, whiting, and sand mortar. The size of the stone
specimens varied in size from 102 mm (4 in) to 300 mm (12 in) cubes. Larger
stones were placed in the bottom portion of the wall.

The back of the wall (northern exposure) has two set-backs, the tops of which are
covered with thin stones forming water shelfs, to study the effects of ponding and
freezing water on the durability of sandstone and marble slabs. Up to the first
water shelf, the back of the wall is covered with concrete except for 23 blocks of
granite, which were set in the concrete as it was poured. Between the first and
second water tables the wall is faced with sandstone and granite. Some of the
stone facing on the back is of thin slabs for testing discoloration effects. The ends
of the wall are faced with sandstone, limestone and granite of various sizes and
shapes. More details of the wall construction have been reported by Kessler and
Anderson [2].

### 2.2. Stone Collection

The wall contains 2,352 individual samples of stone, of which 2059 samples are in
the front (south side) and 293 are in the back and ends. Of these, 2032 are
domestic stone from 47 states, and 320 stones from foreign countries. Over 30
distinct types of stones are represented, some of which are not commonly used for
building purposes. There are many varieties of the common types used in
building, such as marble, limestone, sandstone, and granite. A more detailed

description of the individual stones is given in Table 1 of the report by Kessler and Anderson [2], along with an identification of the specific location of the stone specimens in the wall. Unexposed specimens have been stored indoors at room temperature and humidity to facilitate the determination of weathering effects.

### 2.3. Orientation and Exposure Conditions

The wall faces south (both at the original and new sites). While at the original site the wall was shaded by trees during periods of the day, at the new site it is fully exposed to sunlight. The climate at the stone wall exposure site (both the original and the present site) is characterized as humid continental, typified by warm, humid summers, and short, cool winters [3]. The average monthly temperature varies from around 1 °C (34 °F) in January to 26 °C (78 °F) in July. The maximum temperature exceeds 32 °C (90 °F) for an average of 30 days per year.

Precipitation averages around 970 mm, snowfall around 559 mm; and about 45 freeze-thaw cycles occur per year. The levels of air pollution and acid precipitation are typical of an urban environment. Thus, the stone specimens are exposed to conditions which support several degradation processes including frost attack, thermal and moisture expansion and shrinkage, and acid attack.

### 3.0 Descriptions of Selected Stone and Mortars

The stone's physical conditions range from apparently unaffected to severely deteriorated. The examples provided below represent each of the primary rock types but are not necessarily typical of the condition of all the rocks of that group.

### 3.1 Limestone

The limestones exhibit the widest range in changes in color and texture. The color change is primarily due to the accumulation of organic matter or dirt and soot on the surface. This color change is most apparent on the limestone coping and less prominent in blocks of the same stone that have been set vertically in the wall. Examination of the photograph of the wall a few months after construction (Figure 1) shows some darkening of the coping stone. All of the limestones exhibit an increase in surface roughness, which may serve as a trap for dirt and soot. For example, the Carthage Missouri limestone, known as Carthage Marble, is found as a quoin on the left side of the wall. This stone was described as a gray medium-grained crystalline calcite limestone [2]. Figure 3 illustrates a comparison of the archived core and quoin from the exposure site. The quoin exhibits a striking change in surface texture as the fossil fragments now stand in relief. This change has been remarkably uniform and the weathered stone, while rougher, appears sound.

The Mankato limestones are fine-grained dolomitic limestones that differ in color from buff to gray-buff and cream, and are set as coping stones on both ends and center of the wall, and a single specimen on the East side (Figure 4). This stone exhibits a blackening discoloration and roughening of the surface typical of all the coping stones. The specimen mounted on the East side of the wall exhibits slightly less darkening. Similarly, other limestones mounted vertically appear to have increased surface relief but less darkening.

## 3.2 Marble

Carrara marble is described in [2] as a white, fine-grained calcitic stone with pale-gray clouds. This stone has also undergone deterioration, as the surface texture is sugary with localized, faint iron oxide staining, and rounding of stone edges. The roughness of the surface obscures the gray-cloud coloration of the original surface.

## 3.3 Sandstone

The Triassic red sandstones were popular building stones in the Washington D.C. region and may be found in older commercial, public, and residential construction, including the original Smithsonian building. These red Triassic sandstones originally appear grey and upon exposure weather to shades of red to brown, and have proven to be a easily worked, uniform, and durable building stone [4]. The specimen in the wall is a fine-grained, deep red sandstone and appears to have retained its original smooth surface. Well-defined edges, no scaling, and no apparent discoloration attest to its success as a building stone.

The name Potomac marble is a misnomer as the rock is actually a conglomerate. It was deposited as a series of alluvial fans along the eastern slope of the Blue Ridge Mountains and is comprised of limestone and quartz pebbles and cobbles held together by a carbonate cement [4]. This stone acquired a reputation for being difficult to work but polishes well and can be found as columns in the U.S. Capitol [4]. This stone appears suitable only for protected sites as the sample in the exposure site exhibits significant weathering of both the cobbles and sandy matrix, and has turned from red to grey in coloration.

Figure 3.  Carthage Marble from Missouri, with the archive specimen core on the left, exhibits an increase in surface roughness with fossil fragments standing in relief. The surface is uniform and the edges have retained their sharpness.

Figure 4.  The Mankato Limestone exhibits an increase in surface roughness and blackening common to the other carbonate coping stones. Field width: 25 cm.

Figure 5. The Carrara Marble exhibits a sugary surface texture and surface pitting, localized iron staining, and loss of edge definition and gray-cloud coloration. Field width is 12 cm.

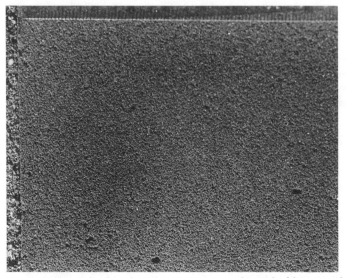

Figure 6. The Triassic red sandstone appears unchanged with a smooth surface, few pits, and sharp edge definition.

Figure 7.  The Potomac marble has performed well in protected locations but
weathers quickly with a color change from red to gray, and visible surface erosion.


3.4  Granite

Two Mount Airy granites from North Carolina presented in Figure 8 show distinct
differences in performance.  The first stone is described as a light-gray, medium-
grained biotite granite  and is placed as a base stone.  The second is a yellowish-
gray, medium-grained biotite granite and is placed in the central part of the wall.
The first stone (Figure 8 A) appears to have performed well and exhibits no pitting
and retains a sharp edge.  The second sample (Figure 8 B) shows substantial
erosion of the biotite (dark) and feldspars (light) with only the quartz (intermediate
gray) retaining the original polish.

3.5  Mortars

Both mortars (Figure 9) have exposed sand grains and are stained with a black
material similar to that staining the coping stones.  The Portland cement mortar
however, appears to be in better condition while the lime mortar exhibits greater
efflorescence and crumbling.  This crumbling may be worse in regions where the
stones have performed poorly.

Figure 8. The North Carolina granites exhibit distinct color differences probably related to their mineralogy and differences in their ability to resist weathering. The light gray stone labeled "Mount Airy" (upper image) was not as affected by weathering. Image field width is about 12 cm.

Figure 9. Both mortars have some cement paste erosion and black staining. The Portland cement mortar (A) exhibits less efflorescence and crumbling than the lime mortar (B). Image field width is about 12 cm.

4.0 Condition Assessment through Image Analysis

As the 50th year of the test wall approaches, plans are being developed for characterizing the stone in the test wall. An image database will be established of both the stone in the wall and the archive specimens through digital imaging. The application of stereophotogrammatic techniques will aid in quantifying surface roughness, and serve as a datum from which to monitor future deterioration. An example of stereo imaging provided in Figure 10 shows the effects of weathering, in the form of erosion along bedding planes, of a Triassic medium-grained brown sandstone from Connecticut. Digital image processing may be used for reconstruction of surface profiles that may be used to monitor the subsequent effects of weathering.

Finally, a petrographic study of the mineralogy and microstructure of the archive stones should serve to provide a basis from which to understand their field performance. These studies may be accompanied by petrographic analyses of micro-cores of the wall specimens. These data should include mineralogy, rock texture, weathering-related changes, color change, change in surface finish, chipping and spalling, and sources and composition of efflorescence.

Figure 10. Stereo pair of a medium-grained, brown Connecticut sandstone shows the erosion due to weathering. Stereophotogrammatic analysis can be used to map the surface profile for assessment of the effects of weathering. Image field width is 31 cm.

5.0 Summary

The National Institute of Standards and Technology Stone Test Wall was originally built in 1948 and was moved intact in 1977 to its present site at NIST in Gaithersburg, MD. The purpose of the stone test wall is to study the performance of stone subjected to weathering. It contains 2,352 individual samples of stone, 2032 of which are domestic stone from 47 states, and 320 stones from foreign countries. Unexposed specimens have been stored indoors at room temperature and humidity facilitating the determination of weathering effects. Almost 50 years of exposure has caused significant changes in some of the stone while others appear almost unaffected. Establishing a digital image database of both the exposed and archived specimens should facilitate future evaluation of the effects of weathering on the stone. Petrographic studies complement the imaging study and should aid in our understanding of the performance of the stone.

6.0 Acknowledgments

The financial support of the Building Materials Division of the Building and Fire Research Laboratory is gratefully acknowledged. The work of those on the stone collections, and those who developed and executed the plans for the stone wall, and who maintained the collection, is also acknowledged.

7.0 References

1. G.W. Hawes, "The Building Stones of the United States and Statistics of the Quarry Industry," *in* Report of the 10th census of the United States, Vol. 10., 1880 399 pp.

2. D.W. Kessler and R.E. Anderson, "Stone Exposure Test Wall," Building Materials and Structures Report 125, 1951, National Bureau of Standards (now available through the National Institute of Standards and Technology).

3. K.A Gutschick and J.R. Clifton, "Durability Study of 14-Year Old Masonry Wallettes," pp. 76-95, Masonry: Past and Present, ASTM STP 589, 1975)

4. Building Stones of our Nation's Capitol, United States Geological Survey INF-74-35, 1975, 44 pp.

# MECHANICAL CHARACTERIZATION OF NATURAL BUILDING STONE

L. Biolzi[1], S. Pedalà[1], and J.F. Labuz[2], Member ASCE

## Abstract

Experiments were performed on a medium-grained granite in order to observe and characterize the zone of localized damage at failure. Geometrically similar three-point bend beams were considered. The size and shape of the localized damage zone appearing at peak load was identified through an interferometric technique (Electronic Speckle Pattern Interferometry) and the locations of acoustic emissions. These measurements are useful in explaining why structures composed of rock can fail at a stress much lower than a strength value determined in the laboratory.

## Introduction

The cultural heritage of many nations consists of a great variety of items and artifacts of high intrinsic value. Such items are often composed of natural building stones. The main natural building stones that have been used in the past to build historical monuments in Europe were granite, limestone, marble and sandstone. The elements (stones) of these structures exhibit inhomogeneity (pores, cracks, grains of various minerals) and frequent anisotropy. Historical monuments usually have been subjected, through time, to mechanical loads, temperature variations, various type of weathering, aging, etc., which affect their quality and impose damage of certain types and degrees. Mechanical damage in natural building stones is manifested by the presence of cracks of various sizes and patterns, developed mostly in the tensile stress regime (e.g. axial splitting cracks, surface spalling, tensile cracks in flexural members, etc.). Knowledge of the mechanical properties and damage diagnoses of stones is the first step in the restoration process. An erroneous diagnosis may be very harmful both to the structural and economic outcome of the operation.

---

[1] Department of Structural Engineering, Politecnico di Milano, Milano, Italy.
[2] Department of Civil Engineering, University of Minnesota, Minneapolis, Minnesota 55455.

Commonly, this evaluation is based only on personal experience and qualitative visual inspection of the exposed stone surface. Less frequently, more elaborate analyses involve sampling and testing, though without any widely accepted methodology. Therefore, one runs the risk of unnecessarily replacing natural building stones that are not damaged to the extent that they appear (by freshly carved and sculptured stones) with the consequence of (a) seriously affecting the integrity of a monument or a part of it, and (b) drastically and often irreversibly altering the historical character of the structure. Moreover, one should keep in mind that visual inspection of an exposed surface does not always reveal internal damage.

A basic concern in restoration projects involving stressed materials is the prediction of failure, where the aim of testing is to provide relevant material properties. A problem, however, arises for brittle materials such as natural building stones because of a size effect. Often times structures composed of rock have been observed to fail at a stress much lower than a strength value determined in the laboratory. In fact, it is well recognized that the nominal strength of natural building stones is dependent on size—as the size of the specimen increases the strength decreases to some limiting value (Bazant and Kazemi 1990).

In this paper, the size effect presented by a medium-grained granite used as a natural building stone is discussed. Geometrically similar three-point bend beams were considered. The size and shape of the localized damage zone were identified through an interferometric technique referred to as electronic speckle pattern interferometry and the locations of acoustic emissions. During the tests, the observation of fringes in real time on a monitor allowed a qualitative monitoring of the crack length at failure.

## Experimental Apparatus

Flexural tests were performed on a granite (Montorfano granite, a white-gray medium-grained rock) with an average grain size of 6 mm and composed of feldspars, quartz and mica. Young's modulus (E) and Poisson's ratio ($v$) in compression were found to be 20 GPa and 0.16, respectively. The specimen geometry was held constant at a span-to-height (L/H) ratio of 5.7. The dimensions of the beams in mm (span x height) were about 2400 x 400, 1200 x 200, 480 x 80, and 240 x 40, with a thickness (T) of 60 or 30 mm. The beams were loaded in three-point bending. The specimens contained a sawn notch, one tenth of the height, to facilitate post-peak control and to provide a predetermined site to observe the damage zone. Experiments were conducted in an Instron closed-loop, servo-hydraulic, 1 MN capacity load frame with the feedback signal taken as the displacement measured across the sawn notch at a gage length of 15 mm. A strain-gage based transducer monitored this so-called crack mouth opening displacement, which was programmed to increase at a rate of $2 \times 10^{-4}$ mm/s. One linear variable differential transformer, an LVDT with a linear range of

0.25 mm, was attached in the centerline of the beam to measure the load-point displacement. Test control and data acquisition were provided by a microcomputer.

## Electronic Speckle Pattern Interferometry (ESPI)

ESPI is an interferometric technique to measure a deformation field of diffusely scattering objects on the surface of the specimen (Jones and Wykes, 1989). In the present tests, the following components were used: a Melles-Griot He-Ne laser (wavelength, $\lambda$ = 632 nm, power, P = 30 mW), a Panasonic WV BP310/G CCD camera, and a DT-2861 frame grabber. In ESPI, by superimposing a reflected light to a reference beam, an interference phenomenon is produced. The resulting interference fringes measure the light path difference between the two beams as multiples of the wavelength, $\lambda$. In this way, by comparing two recorded interference patterns before and after an object displacement, the deformation can be evaluated. The displacement vector field is evaluated by illuminating the specimen from different directions. The image of the object is acquired by a TV camera and transmitted to a monitor through a frame grabber image processing board controlled by a 486 IBM compatible personal computer. Fig. 1 shows a typical recorded image.

Figure 1: Recorded image using ESPI

## Acoustic Emission

The acoustic emission signals generated in laboratory specimens are captured using piezoelectric (PZT-5A) transducers attached to the specimen surface, and preamplified before recording. The sensors have a reasonably flat frequency response from 0.1 to 1 MHz and a sensor diameter of about 3 mm. They are mounted directly to the material with a methyl-cyanoacrylate glue and catalyst. Preamplifiers (40 dB gain) and filters (bandpass from 0.1 to 1.2 MHz) were chosen to maximize amplification, minimize noise, and assure matched frequency response. The data acquisition system consists of four, two-channel digitizers with a sampling rate of 20 million samples per second per channel (50 nanoseconds between two consecutive samples) and 8-bit resolution. The controller interfaces with a personal computer via a GPIB cable and an AT-GPIB card. The digitizers are equipped with an internal trigger that is activated whenever an AE signal exceeds the preset value. This threshold of amplitude must be set so that environmental noise does not trigger the system. The trigger-out signal of the first digitizer is fanned out to the other three digitizers for the acquisition to begin simultaneously at all the digitizers. One of the trigger-out signals is also sent to the load-displacement data acquisition system to correlate AE with the loading history.

In general, source location techniques involve a network of AE sensors positioned at different points on the specimen. Microseismic activity due to a change in stress is detected at each sensor at a given time. By knowing the relative arrival times of the P-wave, which is the component of the signal that arrives first, the P-wave velocity of the material, and the coordinates of each receiver, the event hypocenter can be estimated with a minimum of four sensors. (The problem contains four unknowns: the spatial coordinates of the event and the time at which the event occurred; a fifth sensor is sometimes needed to remove ambiguities arising from the quadratic nature of the distance equation when the sensors are positioned poorly.) Because some error is associated with arrival-time detection (it is not always clear when the signal arrives) and with the P-wave velocity measurement (as damage accumulates material properties may change or become anisotropic), the number of sensors should be increased so that the location problem becomes over-determined. Then, a solution scheme can be developed whereby the error is minimized to obtain a "best fit" type of solution, and statistical methods can be used to evaluate the goodness of the fit (Biolzi and Labuz 1997).

## Experimental Results

An important aspect of the mechanical behavior of natural building stones is that strength is size dependent; that is, material may fail at a quite different stress, usually smaller but possibly larger, than the strength value determined in the laboratory. For beams, the nominal strength $\sigma_n$ of the material, that is, the maximum allowable stress evaluated according to the elementary methods of classical beam theory, can be defined by the ratio M/S, where M is the maximum bending moment and S is the elastic section modulus. Fig. 2 shows the nominal strength as a decreasing function of the size. This is in agreement with other published data (for instance, Bazant and Kazemi, 1990).

From the ESPI fringe patterns, it is possible to identify the end of the localized damage zone as the point where the fringes appear sharply broken (Fig. 1), denoting a discontinuity in the displacement field. The "crack tip" appears at the end of this region. In this way, it is possible to obtain load-crack length curves that, for two different sizes, appear as in Fig. 3. It may be interesting to compare the crack length at the peak load with the corresponding AE locations (Fig. 4). This results in the AE locations appearing more deeply diffused in the beams, denoting a large process zone ahead of the crack tip. It may be observed that by combining the two techniques, one has different, but in a certain sense complementary, information: with the AE locations, the geometrical features of the process zone are identified, whereas with the ESPI fringe patterns it is possible to identify the real crack tip. This information is useful in explaining size effect. For example, if these features are a material characteristic and consequently, are present no matter the size of the structure, then the nominal stress at failure will decrease with size and reach a limiting value (Labuz and Biolzi 1997).

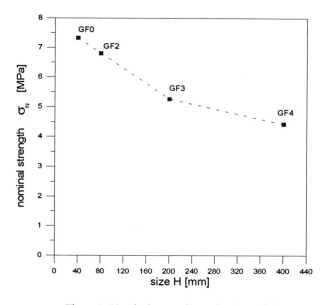

Figure 2: Nominal strength as a function of size

Furthermore, when the size increases, the post-peak response tends to be more brittle. This qualitative behavior of specimens of different sizes may be observed with normalized load-displacement curves (Fig. 5). Dividing the loads and displacements by the corresponding values at the peak load derives a comparison in the overall response. Fig. 5 shows the normalized load-displacement curves for the different beam sizes considered. In the pre-peak branch, the shapes of the curves are similar and no significant differences are observed. Conversely, in the post-peak part of the normalized load-displacement curves, the influence of the specimen size is apparent: the brittleness increases considerably with increasing specimen size.

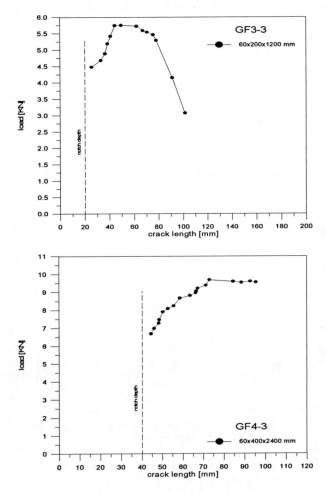

Figure 3: Load-crack length curves

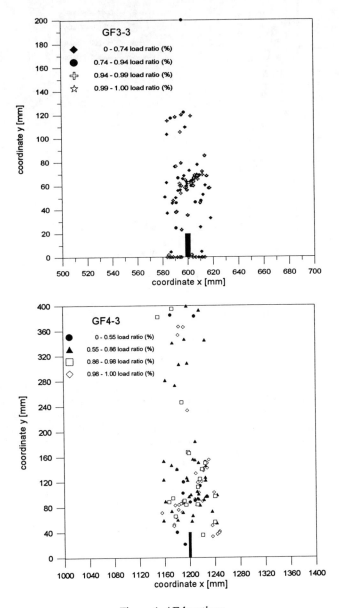

Figure 4: AE locations

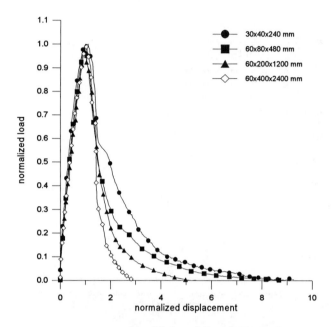

Figure 5: Normalized load-displacement curves

## Conclusions

With the electronic speckle pattern technique (ESPI), it is possible to identify the crack tip at failure as the point where the fringes appear sharply broken. Combining the ESPI measurements and AE locations, two different, but in a certain sense, complementary pieces of information are obtained: with the AE locations, the geometrical features of the process zone are identified, whereas with the ESPI fringe pattern, the crack tip is identified.

## Acknowledgments

Partial support was provided by a North Atlantic Treaty Organization Collaborative Research Grant (no. 950695).

# References

Bazant, Z.P. and Kazemi, M.T. (1990). Determination of fracture energy, process zone length and brittleness number from size effect, with application to rock and concrete. Int. J. Fract., 44, 111-131.

Jones, R. and Wykes, C. (1989). Holographic and Speckle Interferometry, 2nd ed., Cambridge University Press, Cambridge.

Labuz, J.F. and Biolzi, L. (1997). Characteristic strength of quasi-brittle materials. Int. J. Solids Struct. In press.

Constitutive Models for Stone Deterioration

Richard A. Livingston, Associate Member[1] ASCE

Abstract

Stone deterioration processes include a number of agents including salts, air pollution, freeze-thaw and biodeterioration. The resulting distortions of shapes have some analogies with plasticity theory. However, the damage functions used in stone deterioration studies are not the same as those defined for plasticity. Stone damage can be quantified by three different measures: surface recession, mass loss or chemical denudation. Conventional stone damage functions are formulated in terms of environmental exposure variables. In order to develop constitutive models for stone deterioration it is necessary to use environmental loading variables rather than exposures.

Introduction

The objective of this paper is to establish a mathematical formalism for modeling stone damage that is consistent with constitutive models used in strength of materials. Among the advantages of this approach is the ability to make use of existing mathematical theoretical work and the powerful computer codes for handling finite element analysis. In addition, such a framework would be useful in identifying to gaps in the current stone data base and hence the further research required.

In order to understand what is meant by the constitutive model, a brief review of the fundamentals is necessary. The simplest example of a constitutive model is Hooke's Law. In its original uniaxial version it is:

[1]Leader, Exploratory Research Team, Federal Highway Administration, HNR-2, 6300 Georgetown Pike, McLean VA 22101.

$$\sigma = E\epsilon \tag{1}$$

The stress, $\sigma$, giving rise to the strain, $\epsilon$, is independent of the material and involves only the externally applied loads and the geometry of the structure. Conversely, the modulus, E, is a property of the material only, and it is not affected by the geometry of the structure in which the material is placed. It is a constant, although its value can differ from one material to another. This separation of factors into those that are material dependent and those that are loadings, is significant because, as is discussed below, it is not the convention in stone damage research.

The dependent variable, the normative strain, $\epsilon$, is defined as the deformation, $\delta$, divided by the original length L. Thus the deformed length L' is given by :

$$L' = (1 + \epsilon)L \tag{2}$$

or, using Equation (1)

$$L' = \left(1 + \frac{\sigma}{E}\right)L \tag{3}$$

The mathematical form of Equation (3) is that of the class of geometrical affine transformations known as a similitude, and is either a contraction or a dilation, depending on whether $\sigma$ is a compressive or tensile stress. It is a self-similar transformation of L because the factor in the parentheses, the scale factor or ratio of magnification, is a constant (Coxeter, 1969). In physical terms, the effect of the applied stress is to move each point along L a distance that is proportional to the overall length. An important aspect of using the strain rather than the actual deformation is that the relationship given in Equation (1) becomes independent of length scales.

Although it was developed in the context of mechanics of materials, the constitutive model concept need not be limited to mechanical stresses and strains. It can be generalized to other types of loadings and resulting effects such as electromagnetism (de Hoop, 1995). In fact, for stone deterioration, mechanical loads are usually not as significant as other processes that cause stone deterioration. Millennia of experience with construction in stone have taught how to avoid structural failure (Heyman, 1995; Mark, 1982). These processes can include attack by air pollution, freeze-thaw damage and salt weathering. Nevertheless, if these processes can be properly expressed in the form of environmental stresses or loadings, and resulting measures of damage, then equivalent stone-related modulus can be developed.

Before considering in detail its application to stone, some other aspects of the constitutive model approach must be mentioned. In a three-dimensional structure, the stress and strain must be resolved as arrays of normal and shear components. This means that elastic modulus is not a single constant but a matrix of constants. In addition, the matrix of Poisson ratios is required to express the coupling among stresses in one direction and strains in the others.

The moduli are by definition constants. In an ideal brittle material the strain is a linear function of stress up to failure. However, in most materials, the stress-strain relationship becomes nonlinear under certain conditions. This is also true for actual stone, which tends to show some type of inelastic behavior. The exact type of behavior depends on the lithology (Winkler, 1994).

In this situation, the secant modulus of Equation (1) is replaced by the tangent modulus:

$$E_{\tan} = \frac{\Delta\sigma'}{\Delta\epsilon'} \tag{4}$$

where $\sigma'$ and $\epsilon'$ are the coordinates of a point on the inelastic part of the curve. This leads to a definition of damage, d, to the material in terms of a change in stiffness (Lemaitre and Chaboche, 1989):

$$d = \left(1 - \frac{E_{\tan}}{E}\right) \tag{5}$$

Also, the material is usually treated as a continuum so that local variations in properties can be ignored. However, for some heterogeneous materials such as composites, it may not be possible to ignore these fine scale effects and hence a micro-mechanics approach is required. There must also be a limit state that describes the failure condition

Stone Damage functions

Instead of the constitutive model, the stone deterioration research community has typically used the damage function function. The exact definition of the term "damage function" itself is a matter of controversy (Livingston,1997). Nevertheless, it is generally of the form:

$$Y(X,t) = f(X)t \tag{6}$$

where $Y(X, t)$ is the measured damage at time t, and $f(X)$ is a function of some environmental variable such as air pollutant concentration. It should be noted that this damage variable, Y, can be quantified in a number of ways, described below. However, it is not the mechanics of materials damage, d, defined in Equation(5) above.

Equation(6) is actually the cumulative damage function. However, this distinction is generally not considered in the stone damage literature. The damage data is usually normalized to an annual basis. Consequently, what is usually reported in the literature is effectively the damage rate function:

$$\dot{Y} = f(X) \tag{7}$$

Thus the damage rate function, $f(X)$, describes how the damage rate increases or decreases with a change in X, the agent of deterioration.

The approach used to obtain relationships of the form of Equation (7) from cumulative damage data is valid only if it is assumed that $f(X)$ is independent of time. This in turn implies that the cumulative damage, $Y(t)$, is a simple linear function of time. As discussed below, this assumption is necessary because of the limitations of the experimental approach and data interpretation methods typically used in stone deterioration research. However, Livingston and Baer (1990) have proposed that fresh stone would have a damage rate that could significantly differ from that for weathered samples.

A more generally valid approach would be to define the cumulative damage function as the time integral of Equation(7):

$$Y(t) = \int_0^t f(X)\,dt \tag{8}$$

Then it can be seen that $f(X)$ of equation (7) should be replaced by the mean damage rate function $\overline{f(X)}$:

$$\overline{f(X)} = \frac{1}{t}\int_0^t f(X)\,dt \tag{9}$$

This definition does not put any constraints on the time dependence of X. In principle, the mean damage function should be computed from a series of instantaneous value of X taken over the measurement period. In practice, a stone damage rate function based on the annual average value of X is calculated. However, this approach assumes that $f(\overline{X}) = \overline{f(X)}$, which may not be valid.

The major types of deterioration processes are listed in Table I. Temperature-induced effects include permanent deformation caused by expansion or ice formation. Hygric damage refers to expansive effects caused by uptake of atmospheric moisture by the mineral phases in the stone itself (Snethlage and Wendler, 1997). Soluble salts in the pores can also cause damage due to humidity cycles. Wind abrasion removes particles. Rain water running over the surface removes stone by mechanical processes and chemical dissolution. The effect of air pollution, primarily sulfur dioxide, converts calcium carbonate stone to gypsum. Finally, biodeterioration is the process of attack by algae, bacteria, fungus and high order plants growing on the stone surface.

Table I:   SUMMARY OF STONE DETERIORATION PROCESSES

| DAMAGE AGENT | DEFORMATION TYPE | AMBIENT MEASURE | LOADING MEASURE |
|---|---|---|---|
| Thermal | Proportional | Air Temperature | Surface Temperatute |
| Ice | Proportional | Air Temperature | Surface Temperature |
| Hygric | Proportional | Relative Humidity | Surface Moisture |
| Soluble Salt | Proportional | Relative Humidity | Surface Moisture |
| Wind Abrasion | Absolute | Wind Speed | Suction / Drag |
| Water Erosion | Absolute | Rainfall | Runoff |
| Air Pollution | Absolute | Air Concentration | Surface Concentration |
| Biological | Absolute | ? | Biofilm mass? Organic Acids? ADP? |

Stone Damage Measures

The damage variable, Y, can be measured in a number of ways. The three quantitative methods that dominate the stone deterioration literature are: surface recession, mass loss and chemical denudation. Although often treated as being interchangeable, they are not all the same. The major features of each are summarized in Table II.

Table II: Comparison of Stone Damage Measures

| DAMAGE MEASURE | DATA TYPE | SPATIAL DISTRIBUTION | TIME FRAME |
|---|---|---|---|
| Surface Recession | Linear Dimension Change | Point by Point | Decades |
| Mass Loss | Total Mass Loss | Uniform | Months-years |
| Chemical Denudation | Mass of Solutes in Runoff | Uniform | Minutes-Hours |

Surface recession is the distance between the original position of a point on the surface at the start of the measurement interval and its final position at the end of the interval. In the stone literature, this is tacitly assumed to be a one-dimensional measurement. The surface recession is then the movement of a planar surface in a direction normal to its original position. However, as discussed by Livingston (1989), surface recession measurement becomes more difficult to do on a real three-dimensional object, where it must be measured at specific points on the surface.

The surface can be described as an array of points in space. The location of each point is given by a position vector $\vec{\mathbf{r}}$ from some arbitrarily chosen origin. For simplicity, it is assumed that the origin is located within the stone object, and that the surface is convex. The direction of $\vec{\mathbf{r}}$ is not necessarily normal to the surface.

If $\vec{\mathbf{r}}$ is the original position vector and $\vec{\mathbf{r}}'$ is the final position vector of a point, then the surface recession, $\delta$, is the magnitude of the difference vector (Fig.1 ):

$$\delta = |\vec{\mathbf{r}} - \vec{\mathbf{r}}'| = \sqrt{r^2 + r'^2 - 2rr'\cos\theta} \tag{10}$$

The surface recession can be resolved into two components:$\delta_\parallel$ and $\delta_\perp$, which are respectively the recession along the original direction $\vec{r}$ and the recession perpendicular to it:

$$\delta^2 = \delta_\parallel^2 + \delta_\perp^2$$

where:

$$\delta_\parallel = r - r' \cos\theta$$

$$\delta_\perp = r' \sin\theta$$

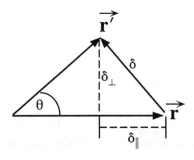

**FIGURE 1: Diagram of Surface Recession**

If $\theta = 0$, then

$$\delta = r - r' = r(1 - a) \tag{10}$$

where $a = r'/r$. which is simply a contraction in the direction along the direction $\vec{r}$. By comparison with Equation (2) it can be seen that $a = 1 + \epsilon$, so that a is the stretch ratio. In this case then, $\delta = \delta_\parallel$ and $\delta_\perp = 0$. If $\theta \neq 0$, then $\delta_\perp$ is nonzero also. This represents a deviatoric component of the strain which changes the direction, and thus a tendency for the overall shape of the object to distort.

In the first case, if a is constant for all $\vec{r}$, that is, the case of uniform contraction, then the object does not change shape; it simply becomes smaller. If a is not uniform, even though there is no deviatoric component then there will still be a distortion of shape (Khan and Huang, 1995).

This far, the surface recession can be regarded as consistent with elastic-plastic deformation in an Eulerian coordinate system. However, the analogy is not complete, because in stone deterioration mass is not conserved. Consequently, it is not possible to label a point on the surface and then measure its displacement with time. Instead, one would first have to establish a point on the original surface, which then fixes the position vector $\vec{r}$. Subsequently, one locates the point on the new surface by its intersection with $\vec{r}$. However, the usual practice is measure the change along a normal to the existing surface. Thus the surface recession measurement is a one-dimensional approximation of a more complicated deformation.

This may not be a significant problem if the local radii of curvature of the surface are much, much larger than $\delta$, so that it appears to be planar on this length scale. However, on a stone sculpture, the details of interest consist of highly curved surfaces on fine length scales. Consequently, the deterioration of the sculpture is perceived as a distortion, not a uniform contraction.

Another consequence of the fact that stone deterioration is a process of mass loss is that it may not be appropriate to express dimension changes in terms of strains. In processes such as water erosion, a certain amount of material is removed per unit surface but the interior is not affected. Moreover, the amount of material removed, and hence the local deformation, is determined by external factors and is not a function of the size of the object. In terms of Equation 10, in this situation, the absolute deformation, $\delta$, is fixed, which means that for different values of $r$ there will be different values of $a$, and hence differing values of $\epsilon$ for the same loading. On the other hand, in the case of processes like thermal expansion, $\epsilon$ would be uniform for a given temperature change, while $\delta$ would vary from point to point. Consequently, as shown in the second column in Table I, it is possible to classify the processes of deterioration in terms of the relevant type of deformation.

Finally, another drawback of the surface recession variable is the relatively low velocities observed, in the range of $10 - 100 \, \mu m/yr$. Given the typical surface roughness of stone, statistically significant surface recessions are not found over time intervals of less than 10 years (Trudgill et al., 1989).

Since the stone damage is a mass loss process, a more direct measure could be the amount of mass lost over a time interval. However, it is difficult to apply this to actual stone objects. In principle, the mass loss can be measured for the entire object. Given the surface area of the object, an average mass loss per unit surface can be calculated. However, weighing a large stone object may not be feasible at all. Mass loss measurements are typically made on small cubes or slabs with a relatively high surface to volume ratio (Haynie, 1983; Webb et al., 1992). In one case, the actual number of grains lost as a function of time was counted from a series of photographs (Dolske et al. 1991). This was then converted into a mass loss by assuming that all grains had the same mass.

The third measure, chemical denudation, measures the amount of solutes such as calcium, dissolved from the surface in rainfall runoff. It is very sensitive with typical values, expressed as equivalent mass loss, of around 30 mg/m$^2$ (Livingston et al., 1982). It is also possible to analyze the data to determine the relative importance of the stone deterioration processes of acid rain neutralization, dry deposition of $SO_2$ and natural dissolution ( Livingston, 1992). Nevertheless, chemical denudation measures only one component of the overall loss of material.

Various authors have attempted to convert the chemical denudation rate measured on carbonate stone into a surface recession rate by dividing it by the density of calcite. This in effect assumes that the stone is a pure crystal of calcite that weathers uniformly. In reality, because of the porosity of the stone, the actual density may be considerably less that than of the stone, and the actual surface area available for dissolution may be much larger than the geometric surface area. Moreover, the lithology of the stone may result in a larger volume material being lost than predicted by the calcite dissolution model. Therefore, it is incorrect to use the term "surface recession" to refer to this length change which is calculated rather than actually measured. The accuracy of the pseudo-surface recession rate could be determined for a given lithology by conducting experiments

There are other damage variables that could be measured. Surface roughness or porosity may provide more sensitive measurements especially during the early exposure times. However, there are not yet standardized methods for making these measurements in the field on actual stone objects.

The damage function thus describes the relationship of those variables listed in Table II on the factors given in Table I. However, the latter are usually stated in terms of exposure variables such as air temperature or air pollution concentration listed in the third column of Table I. The rate of attack, however, depends on the loading, the amount of heat, air pollutant, etc actually that reaches the surface. The relationship between the exposure and the loading may be non-linear and may also involve other variables.

## The Constitutive Model for Sulfur Dioxide Air Pollution

For example, the deposition function for sulfur dioxide air pollution can include as many as eight parameters including ambient air concentration, relative humidity and wind velocity, as well as properties of the stone itself (Livingston, 1997). Nevertheless, it is the convention to treat the air pollution deposition process as a simple linear relationship, i.e.:

$$F = v_d \overline{C_{SO_2}} \tag{10}$$

where F is the flux or the mass of sulfate deposited on the surface per unit of time, $\overline{C_{SO_2}}$ is the annual average atmospheric concentration of sulfur dioxide, and

$v_d$ is a mass transfer coefficient known as the deposition velocity (Chamberlain, 1960). This approach follows from the typical method for studying stone deterioration, which consists of setting out stone samples at a site where the air pollution concentrations are measured using standard instruments. The damage data are then related to the air pollution data using regression analysis (Kucera et al., 1995).

It is further assumed under the "stoichiometric model" (Haynie, 1983) that the stone damage rate in mass lost per unit surface is simply proportional to the $SO_2$ flux, hence:

$$\overline{f(C_{SO_2})} = \frac{M_{calcite}}{M_{SO_4}}F = \frac{M_{calcite}}{M_{SO_4}}v_d\overline{C_{SO_2}} \tag{11}$$

where $M_{calcite}$ and $M_{SO_4}$ are the molecular weights. This states that the deposition of one mole of sulfate produces an amount of stone damage equal to exactly the mass of one mole of calcite. As noted above under the discussion of surface recession and chemical denudation, there is usually not a one-to-one correspondence between calcite mass loss and stone damage rate.

Another problem with the stoichiometric model of Equation (11) is that $v_d$ includes environmental factors such as RH that affect the deposition of the $SO_2$ as well as properties of the stone itself. If the deposition rate, F, could be measured directly, then one could simply write:

$$f_{SO_2} = g(F) \tag{12}$$

Here the symbol $f_{SO_2}$ means the damage caused by the agent sulfur dioxide as opposed damage due to other types of attack. This is in contrast to Equation (11) in which $f(C_{SO_2})$ indicates the damage associated with a given concentration of sulfur dioxide.

Ideally, the stone damage rate would have a linear relationship with the deposition rate. This would be written as:

$$f_{SO_2} = cF \tag{13}$$

The coefficient c is a susceptibility factor that depends only on the properties of the stone. Thus Equation (13) is a constitutive model formulation.

The direct measurement of the deposition rate to a surface would involve scraping the surface off to a specified depth and analyzing the scrapings for the sulfate content (Girardet and Felix, 1981). An indirect approach would be to use standard stone specimens as surrogate surfaces (Luckat and Zallmanizig, 1985; Furlan and Girardet, 1992). Nondestructive measurements of deposition may also be possible using portable neutron gamma, X-ray fluorescence or infrared systems.

Given a method for measuring the deposition rate, the stone susceptibility factor, c, can then be determined. It is likely that this will vary considerably from one type of stone to another, depending on such factors as mineral assemblage and pore-size distributions (Livingston, 1988).

If the relationship between the damage and the deposition is non-linear, then it would be necessary to rewrite Equation **(13)** in differential form as:

$$df_{SO_2} = \frac{\partial f_{SO_2}}{\partial F} dF \tag{14}$$

## Definitions Of Loadings

The key to developing the constitutive model for sulfur dioxide air pollution damage lies in converting from an exposure variable, $C_{SO_2}$, to a loading variable, the sulfate deposition rate. This approach can be generalized by writing Equation **(7)** as:

$$\dot{Y} = \frac{\partial c_i}{\partial Z} dZ \tag{15}$$

where Z is the loading variable associated with the exposure variable X, and $c_i$ is the stone susceptibility factor for the damage process.

Although Z is a function of X, the relationship may be complex and typically involves other site-specific variables. Consequently, Z would usually have to be measured directly. Table I lists exposure variables and possible loading variables for the stone deterioration processes. For example, thermal expansion is determined by the actual temperature of the stone, but stone durability indices have been developed on the basis of air temperature (Fookes et al., 1988). There could be a significant difference between the two temperatures, depending on solar radiation, thermal mass, wind speed etc.

Direct measurement of stone temperature could be made by a thermometer in contact with the surface, or remotely by infra-red sensors. Measurement of some of the other loading variables listed in Table I would require some ingenuity. For example, wind abrasion depends on the forces exerted by the airflow over the surface. The most direct measurement would be by drag force plates embedded in the surface. However, this may not be possible on stone monuments. An indirect method might be to measure the rate of evaporation of some volatile tracer compound applied to the surface. Additional research is required for development of methods for measuring these loading variables. This is particularly true for biodeterioration. There does not even appear to be an exposure variable available that could be used to estimate the biological activity on the surface. Several possibilities for loading variables are given in Table I, but very little has been done to develop them.

Conclusions

The constitutive model approach is useful as a framework for investigating stone deterioration. The crucial feature is the mathematical formalism that explicitly separates the relationship into terms of loading of the agent of deterioration, and the susceptibility of the stone to the loading. These susceptibilities are then materials properties, independent of the exposure variable.

To make use of this approach it is necessary to develop physical methods for measure environmental variables as loadings to the stone rather than exposure variables. Methods for measuring the damage to stone need to be improved as well. Once these methods are available and standardized, it will then be possible to obtain data on the stone's susceptibility to damage as a material property. Ultimately, the susceptibilities may be determined from petrological and geochemical principles rather than empirically.

References

Chamberlain, A.C. (1960) "Aspects of the deposition of radioactive and other gases and particles" *Aerodynamic Capture of Particles*, Richardson, E.G., ed., Pergamon Press, Oxford, 63-88.

Coxeter, H.M.S. (1969) *Introduction to Geometry*, John Wiley & Sons, New York.

Dolske, D., Petuskey, W. and Richardson, D. (1991) "Impascts of microclimate and air quality on sandstone masonry of Anasazi dwelling ruins at Mesa Verde National Park", in: *Proceedings of the Anasazi Symposium 1991*, Hutchinson, A. and Smith, J., eds. Mesa Verde Museum Association, Mesa Verde CO 241.

de Hoop, Adrianus T.(1995). *Handbook of Radiation and Scattering of Waves*, Academic Press, San Diego.

Fookes, P.G., Gourley, C.S. and Ohikere, C. (1988) "Rock weathering in engineering time." *Quarterly Journal of Engineering Geology*, 21, 33-57.

Furlan V. and Girardet, F. (1992) "Pollution Atmosphérique et Réactivité des Pierres." *Proceedings, 7th International Congress on Deterioration and Conservation of Stone*. Delgado Rodrigues, J., ed., Laboratorio Nacional de Engenharia Civil, Lisbon, 153-163.

Girardet, F. and C. Felix (1981) "Accumulation des composes soufres dans un mur en molasse abrite de la pluie a l'eglise St. Francois-Lausanne-Suisse." *The Conservation of Stone II*, R. Rossi-Manaresi ed., Centro per la Conservazione delle Sculture all'aperto, Bologna, 91-106.

Haynie, F.H. (1983) Deterioration of marble, *Durability of Building Materials*, **1** 241-254.

Herman, J. (1995) *The Stone Skeleton: Structural Engineering of Masonry Architecture*, Cambridge University Press, New York.

Khan, A.S. and Huang, S. (1995) *Continuum Theory of Plasticity*, John Wiley & Sons,Inc. , New York.

Kucera, V., Tidblad, J., Henriksen, J., Bartonova, A. and Mikhailov, A. (1995) *Statistical Analysi of 4 year Materials Exposure and Acceptable Deterioration and Pollution Levels,* Swedish Corrosion Institute, Stockholm.

Lemaitre, J. and Chaboche, J.L.( 1989) *Mechanics of Solid Materials*, Cambridge University Press, NY.

Livingston, R. and Baer, N. (1990): "Use of tombstones in the investigation of deterioration of stone monuments", *Environmental Geology and Water Science*, **16** [1] 83-90.

Livingston, R. A. (1988): "the application of petrology to the prediction of stone durability", *VIth International Congress on Stone Conservation*, J. Ciebach ed., Nicholas Copernicus University, Torunn, Poland, 432.

Livingston, R., Kantz, M. and Dorsheimer, J. (1982) *Stone Deterioration at Bowling Green Custom House, 1980-81, Interim Report, EPA 600/6-84-003*, National Technical Information Service, Springfield VA.

Livingston, R.A. (1989) "The computational geometry of the deterioration of stone sculpture." *Structural Repair and Maintenance of Historical Buildings.* C. Brebbia. ed., Computational Mechanics Publications, Southampton, UK, 593-602.

Livingston, R.A. (1997) "Development of Air Pollution Damage Functions" *Saving our Architectural Heritage: The Conservation of Historic Stone Structures*, N.S Baer and R. Snethlage eds., John Wiley & Sons, NY, 37-62.

Livingston, R.A.( 1992) "Graphical methods for examining the effects of acid rain and sulfur dioxide on carbonate stones." *7th International Congress on Deterioration and Conservation of Stone,* J. Delgado-Rodrigues ed., Laboratorio Nacional de Engenharia Civil, Lisbon. 375-386.

Luckat, S. and J. Zallmanzig (1985) *Investigations on the rates of immission and effects in selected places of Europe for the quantitative examination of the influence of air pollution on the destruction of ashlar, Part A. Research Report No 106 08 006,* Zollern Institute of the Deutsches Bergbaumuseum, Zollen, Germany.

Mark, R. (1982) *Experiments in Gothic Structure*, MIT Press, Cambridge MA.

Snethlage, R. and Wendler, E. (1997) "Moisture Cycles and Sandstone Deterioration", *Saving Our Architectural Heritage: The Conservation of Historic Stone Monuments*, N.S. Baer and R Snethlage, eds., John Wiley & Sons, NY, 7-24.

Trudgill, S.T., Viles, H.A., Inkpen, R.J, and Cooke R.U. (1989) "Remeasurement of weathering rates, St. Paul's Cathedral, London." *Earth Surface Processes and Landforms,* 14, 175-196.

Webb, A.H., Bawden, R.J., Busby, A.K., and Hopkins, J.N. (1992) Studies on the effects of air pollution on limestone degradation in Great Britain. *Atmospheric Environment.* **26B** [2]. 165-181.

Winkler, E. (1994) *Stone in Architecture: Properties, Durability,* Springer-Verlag Press, NY.

# NATURAL STONE DESCRIBED AS A MICROCRACKED SOLID

F. C. S. Carvalho[1], J. F. Labuz[1], Member ASCE, and L. Biolzi[2]

## Abstract

Experiments were performed on Charcoal granite to investigate the possibility of measuring the amount of damage present in the material. The results of 2D tensile tests on artificially cracked aluminum plates suggest that, if a random distribution of microcracks in a material is assumed, then the problem can be modeled using the approximation of non-interacting cracks even when interactions occur. The microcrack density of the virgin rock was estimated with the use of pressure-strain analysis, where linear strains were measured as a function of applied hydrostatic pressures. At large enough pressure, typically 50 MPa, all cracks closed and the measured response was due to the solids only. After transformation of the strain data, the volume strain was determined and the modulus of the solids was estimated to be 84 GPa (for Poisson's ratio of 0.25). The modulus of the granite in uniaxial tension was determined to be 46.5 GPa, so the granite in its natural state has a microcracked density of about 0.45. Further damage was then induced in the rock by slow, uniform heating. Results and implications of the additional microcracking are discussed.

## Introduction

Natural building stone is generally thought of as a solid material with substantial stiffness and compressive strength. What is sometimes forgotten or not fully appreciated, however, is that rock is a damaged material: small cracks and pores are common features found, in varying amounts, in all natural stone. Furthermore, measurements of mechanical properties are affected by the type of damage. For example, a microcracked solid can exhibit double elasticity – the modulus from uniaxial compression and tension tests can be different – while a solid with

---

[1] Department of Civil Engineering, University of Minnesota, Minneapolis, Minnesota 55455.
[2] Department of Structural Engineering, Politecnico di Milano, Milano, Italy.

pore-like damage is not affected by the sign of the applied stress. Of course, the presence of damage severely affects the tensile strength of rock, such that the maximum uniaxial stress applied in tension can be 10–20 times smaller than that in compression. In addition, an increase in damage may degrade mechanical properties and cause visible distress to the stone. Thus, a quantitative assessment of damage may be useful in evaluating the state of the material. Various researchers have described theoretically the change in the modulus of an elastic material as a function of crack density. Some of the theories are evaluated by comparing their results with 2D tests on artificially cracked aluminum plates subjected to uniaxial tensile stress.

Most rocks selected as natural stone building facades or historical monuments are subjected to thermal or mechanical loading. The level of loading will dictate if additional damage, mainly in the form of microcracks, will be generated. Experiments were performed on Charcoal granite to assess the amount of damage, as measured by the microcrack density, present in the virgin rock. Because microcracks are very difficult to observe in rock, a technique proposed by Brace and co-workers in the 1960's was used. Essentially, linear strain in at least two directions on three perpendicular planes was measured as a function of hydrostatic pressure. At high enough pressure, all microcracks close and the measured response is attributed to the solid part. The effect of thermal loading on the rock was then investigated by subjecting the specimens to slow, uniform heating and using the pressure-strain technique.

## Effective modulus of 2D cracked solids

### Analytical methods

Several analytical methods have been proposed to describe the effective elastic properties of solid materials containing an arbitrary distribution of interacting or non-interacting cracks. The occurrence of interactions between cracks in a material depends on the relative positions between them. The approximation of non-interacting cracks corresponds to considering each crack as an isolated one, not taking into account the effect of interactions between cracks. This can be done when cracks are relatively far apart from each other, such that each crack does not feel the presence of the others and behaves exactly like one single crack in the material. In this context, mutual positions between cracks do not matter and the solution for the problem can be obtained simply by taking the solution for one single crack in a solid material and making the summation over orientations. In the case of a random distribution of non-interacting cracks, the solution is given by (Kachanov, 1992)

$$E = \frac{E_0}{1 + \pi\rho}; \qquad \nu = \frac{\nu_0}{1 + \pi\rho} \tag{1}$$

where $E_0$ and $\nu_0$ are the elastic moduli of the matrix material, $E$ and $\nu$ are the effective elastic moduli of the cracked material, and $\rho = \frac{1}{A}\sum a^2$ is the scalar crack density parameter ,which is a function of the individual crack lengths $2a$ and a representative area $A$.

The determination of effective elastic properties of cracked materials becomes significantly more complex when interactions between cracks occur. This is due to the fact that the solution to the problem, for each realization of crack statistics, would have to include information not only on the number of cracks, sizes and orientations, but also on mutual positions between them. Due to the difficulty in processing such calculations, some approximate schemes have been proposed.

The self-consistent scheme (Budiansky and O'Connell, 1976) and the differential scheme (Hashin, 1988) both consider each crack as an isolated one in the matrix with reduced stiffness, with the only difference being that, in the case of the differential scheme, the analysis is done incrementally. In the case of randomly located cracks, however, their predictions for the ratio $E/E_0$ are very different ($1 - \pi\rho$ for the self-consistent scheme and $e^{-\pi\rho}$ for the differential scheme). Another popular approximate scheme, the method of Mori-Tanaka (Mori and Tanaka, 1973), consists of placing a representative crack into the undamaged matrix and subjecting it to a homogeneous effective stress field. If cracks are randomly located, the Mori-Tanaka's predictions coincide with the approximation of non-interacting cracks.

## Experiments

The effective modulus of a 2D solid containing cracks was obtained by subjecting an artificially cracked elastic material to a uniaxial tensile stress under plane stress conditions. The specimen used was a rectangular aluminum (alloy 2024) plate, $1.6mm$ thick, with a plane area of $685.8mm \times 228.6mm$. Two LVDTs, one at each side of the plate, were used to measure the vertical displacements corresponding to its central part, where 20 slots were randomly located (Figure 1). The value of strain used for calculation of the effective Young's modulus was obtained from averaging the responses of the two LVDTs and dividing by an appropriate gage length (Carvalho and Labuz, 1996). The stresses were obtained by dividing the applied loads by the cross-sectional area of the plate ($365.8mm^2$). Dividing the axial stresses by the average axial strains yields the effective Young's modulus of the inhomogeneous material.

Slots were used instead of cracks due to the difficulty in creating real cracks and were cut through the thickness of the plate to characterize the plane stress condition. The lengths of the slots were calculated such that a minimum aspect ratio of 1 : 10 was obtained, in order to have crack-like features. For this aspect ratio, the difference in the compressibility of the slots as compared to the case of cracks is only 1% (Kachanov, 1992).

Figure 1: Plate layout for crack density of 0.09

A random number generator was used to create the positions of the centers of the slots over the central third of the plate, without intersecting each other. In order to assure that an approximate isotropic distribution would be obtained, and because the dimensions of the plate presented a limitation on the number of slots that could be used, their orientations were not random, but were prescribed to uniformly vary from 0 to 180 degrees. To ensure a representative behavior of the nonhomogeneous element, slots were also located outside the central part by reflection of the positions already existing. The plate was tested for values of crack density between 0.02 and 0.15, and for each value all the slots in the plate had the same length .

The results are presented in terms of a ratio between the effective Young's modulus $(E)$ and the Young's modulus of the intact material $(E_0)$ as a function of crack density (Figure 2). The experimental results were compared to some of the existing approximate schemes for the case of interacting cracks. The topmost line in the graph represents the analytical results obtained from the approximation of non-interacting cracks, which coincides with the predictions of Mori-Tanaka's scheme. The other two lines in the graph correspond to the predictions of the differential scheme and the self-consistent scheme. The experimental results are in very good agreement with the non-interacting theory even at higher values of crack density, where interactions probably occur. The differential scheme predictions are also close to the experimental results for lower values of crack density. However, at higher densities the non-interacting approximation provides better predictions. The experimental results are not in good agreement with the predictions of the self-consistent scheme, which tends to overestimate the effective compliance.

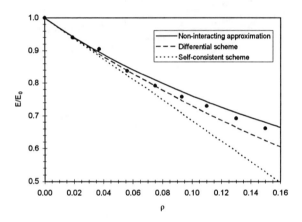

Figure 2: Effective Young's modulus of a 2D plate with random slots

## Thermally induced microcracks in rock

### Pressure-strain analysis

The effect of thermal loading on the effective moduli of Charcoal granite specimens was investigated with the use of a technique called pressure-strain analysis. This technique was first applied to the characterization of damage in rocks by Simmons et al. (1974), and consists of measuring linear strains on three perpendicular planes of the specimen as a function of the applied hydrostatic pressure. As pressure increases, the existing microcracks in the material start closing. At high enough values of pressure, typically 50 MPa for the tests performed, all microcracks were closed and the measured linear response can be attributed to the solids only.

Figure 3 shows the behavior of a virgin Charcoal granite specimen subjected to pressures up to 140 MPa. The specimen shows anisotropic behavior at the beginning of the test, before all microcracks were closed, and an approximately isotropic behavior later on. This suggests that the matrix material (the solid part) is actually isotropic and that the earlier observed anisotropy could be explained by the progressive closing of microcracks, also reinforcing the idea that damage is present in rocks even in their virgin state.

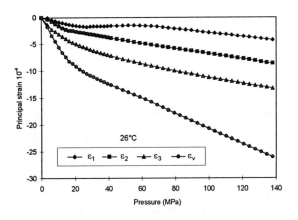

Figure 3: Pressure-strain response of a virgin granite specimen

The effect of thermal loading on the behavior of rocks was then investigated by further damaging the same material through slow, uniform heating. The specimens were subjected to a thermal gradient of 2°C/min until the desired temperature of 300°C was reached, and was held constant for 4 hours. After reaching again room temperature, the specimens were tested following the same technique as before. The results in Figure 4 show a similar behavior as compared to the previous test. The material also presents a non-linear anisotropic behavior for pressures up to about 50 MPa, but the magnitude of strains has now increased.

A simple way of comparing the amount of damage present in both specimens is by looking at the y-intercept of the linear portion of the volumetric strain curve. Since the linear behavior is attributed to the response of the solid part only, that value can be interpreted as the volumetric strain produced upon

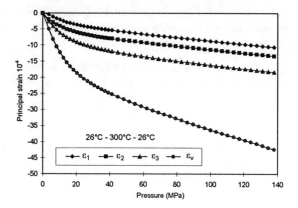

Figure 4: Pressure-strain response of a granite specimen preheated to 300°C

closure of the microcracks. The thermal loads have caused an increase in the microcrack-closure strains from 670 to about 1,900 microstrain, which suggests a substantial increase in the amount of damage.

## Estimate of microcrack density

In order to obtain an estimate of the microcrack density of the granite, in both virgin and damaged states, a few assumptions had to be made regarding the type of damage present in the rock, for example in the form of microcracks or pores. Because Charcoal granite is a rock of very low porosity, less than 1%, the effect of pores on the behavior of the rock was neglected and it was assumed that damage is present in the form of microcracks only. For further simplification, and because this rock is considered to be isotropic, it was assumed that a random distribution of penny shaped cracks in an isotropic matrix could provide a reasonable representation of the damaged rock. As demonstrated by the 2D experiments, the effective moduli in this case can be obtained from the approximation of non-interacting cracks, which yields the following equations for 3D solids (Kachanov, 1992):

$$\frac{E}{E_0} = \left[1 + \frac{16(1 - \nu_0^2)(1 - 3\nu_0/10)}{9(1 - \nu_0/2)}\rho\right]^{-1} ; \qquad \frac{\nu}{\nu_0} = \frac{E}{E_0}\left[1 + \frac{8(1 - \nu_0^2)}{45(1 - \nu_0/2)}\rho\right] \tag{2}$$

In the above equations, $E_0$ and $\nu_0$ are the elastic moduli of the matrix material, $E$ and $\nu$ are the effective elastic moduli of the cracked material, and $\rho = \frac{1}{V}\sum a^3$ is the scalar crack density parameter which is a function of the individual crack diameters $2a$.

The bulk modulus of the intact rock can be estimated from the relation between pressure and volumetric strain for a linearly elastic material:

$$p = -k_0\,\varepsilon_v \tag{3}$$

where $k_0$ is the bulk modulus. By taking the slope of the linear portion of the $p - \varepsilon_v$ curve in the graphs, the average value obtained is $k_0 = 56.06\ GPa$, and it is related to the elastic properties of the matrix material by

$$k_0 = \frac{E_0}{3(1 - 2\nu_0)} \tag{4}$$

Assuming a value of Poisson's ratio of 0.25 (Love, 1944), the elastic modulus of the solid part calculated is $E_0 = 84.1\ GPa$. The effective Young's modulus of the virgin and thermally damaged materials were obtained from flexure tests, which yielded $E = 46.5\ GPa$ for the natural granite and $E = 27.1\ GPa$ for the granite heated at 300°C. These values were inserted into equation (2) to give an estimate of the corresponding crack densities. The results suggest that the applied thermal loading has caused an initially calculated value of crack density of 0.45 to increase to about 1.19, resulting in a reduction of the Young's modulus of about 50%.

## Conclusions

Some of the proposed analytical methods for calculating effective properties of cracked materials were evaluated by comparing their predictions to 2D tests in artificially cracked aluminum plates subjected to uniaxial tensile stresses. The results obtained showed that, in the case of randomly located cracks, the approximation of non-interacting cracks provides a very good estimate of the effective Young's modulus. The differential scheme also provides a good approximation to the problem at low values of crack density, while the self-consistent scheme significantly overestimated the effective compliance for the case studied.

The effects of thermal loading on the amount of damage present in Charcoal granite specimens were estimated by using the pressure-strain analysis technique. The tests performed with the virgin rock and the rock damaged at 300°C showed a significant increase in the amount of volumetric strains corresponding to microcrack closure. The crack densities for both cases were then estimated by assuming that the damage in the material was in the form of randomly distributed penny-shaped cracks which, as shown by the 2D tests, can be modeled with the approximation of non-interacting cracks. The applied thermal loadings resulted in a reduction of the effective Young's modulus of the granite of about 50%, due to an increase in microcrack density from 0.45 to 1.19.

## Acknowledgments

Partial support was provided by a National Science Foundation Grant (no. CMS–9532061) and a North Atlantic Treaty Organization Collaborative Research Grant (no. 950695).

## References

[1] Budiansky, B. and O'Connell, R. J. (1976). Elastic moduli of a cracked solid. *Int. J. Solids Structures*, **12**, 81–97.

[2] Carvalho, F. C. S. and Labuz, J. F. (1996). Experiments on effective elastic modulus of two-dimensional solids with cracks and holes. *Int. J. Solids Structures*, **33(28)**, 4119–4130.

[3] Hashin, Z. (1988). The differential scheme and its application to cracked materials. *J. Mech. Phys. Solids*, **36**, 719–734.

[4] Kachanov, M. (1992). Effective elastic properties of cracked solids: critical review of some basic concepts. *Appl. Mech. Rev.*, **45(8)**, 304–335.

[5] Love, A. E. (1944). *A Treatise on the Mathematical Theory of Elasticity*. Dover Publications, New York.

[6] Mori, T. and Tanaka, K. (1973). Average stress in matrix and average elastic energy of materials with misfitting inclusions. *Acta Met.*, **21**, 571–574.

[7] Simmons, G., Siegfried, R. W., and Feves, M. (1974). Differential strain analysis: a new method for examining cracks in rocks. *J. Geophys. Res.*, **79**, 4383–4385.

# Evaluation of Stone Cladding Anchorages
## on Precast Concrete Building Panels

Kurt R. Hoigard[1] and George R. Mulholland[1]
ASCE Associate Members

## ABSTRACT

Due to reported cracking, displacement, and emergency removal of 25 stone cladding pieces on an 18 story Federal office building, the authors were asked to evaluate the stability of the facade, determine the cause of the observed behavior, design repairs to address deficiencies revealed by the investigation, and provide quality assurance monitoring of the repair construction. This paper focuses on the laboratory, analytical, and field investigative methods used to evaluate in-place integrity of the stone cladding anchorages.

## INTRODUCTION

Stone clad precast concrete building panels have been in common use in the U.S. since the 1960s, and were incorporated into a large number of buildings during the construction boom of the mid to late 1980s. Buildings enclosed with this system enjoy the classic appearance of stone, combined with the ease and speed of erection inherent in prefabricated panelized construction. With most products, advantages are balanced by disadvantages, and stone clad precast panels are no exception. Disadvantages associated with this cladding system include:

1.  Special measures must be taken at the factory with the stone, concrete, and stone anchors to ensure sound anchorage of the thin stone veneer to the concrete backing structure.

2.  Care in handling must be exercised during storage, transportation, and erection to avoid damaging the cladding stones.

---

[1] Principal and Project Engineer II, respectively, Raths, Raths & Johnson, Inc., 835 Midway Drive, Willowbrook, IL 60521-5591

3.      Damaged or improperly anchored stones are difficult and costly to repair or replace after the precast panels are erected.

This paper discusses recent investigative experience by the authors, involving a stone clad precast concrete building facade with stone anchorage deficiencies attributed to improper fabrication, installation, and field repair procedures. Laboratory, analytical, and field investigation methods used to evaluate in-place integrity of the stone cladding anchorages are presented.

## PROJECT DESCRIPTION

The subject building is an 18 story office building located in the Washington D.C. area, and owned by the real estate arm of the federal government known as the General Services Administration (GSA). The exterior facade consists of 10.2 cm (4 inch) thick stone clad precast concrete panels with aluminum framed fixed windows fitted into punched openings in the precast. The majority of the cladding stone is 2.5 cm (1 inch) thick white marble with a honed finish. Polished 2.2 cm (7/8 inch) and 3.2 cm (1-1/4 inch) thick mahogany colored granite was reserved for the main entrance elevation and penthouse areas (refer to Figures 1 and 2).

Figure 1. Overall View of Building

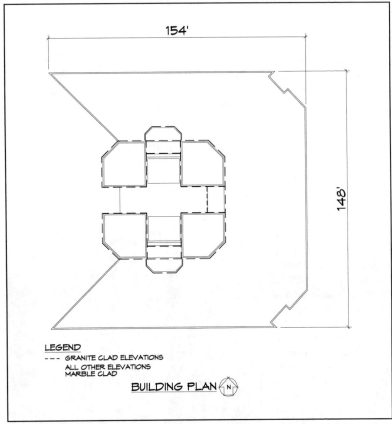

Figure 2. Building Plan

Over 14,000 pieces of stone were used on the project, encompassing a total of 48 combinations of stone type and size (refer to Table 1). Multiple 4.8 mm (3/16 inch) diameter hairpin anchors of the type shown in Figure 3 were used to attach the stone pieces to the precast concrete back-up. As shown in the figure, the "legs" of the anchors fit into holes drilled into the backs of the stone pieces, with the remainder being cast into the concrete portion of the panel. Panel fabrication was accomplished by laying the stone pieces face down into an appropriately sized form, placing duct tape across joints between stones to prevent concrete leakage, spraying a liquid-applied bond breaker onto the backs of the stones, inserting the anchors into the pre-drilled holes in the backs of the stones, arranging the steel reinforcing bars required for the concrete portion of the panel, and pouring the concrete into the mold. Approximately 16 hours after the concrete pour, the panels are removed from the molds and set aside to cure.

## Table 1: Stone Sizes and Corresponding Number of Anchors

| STONE SIZE (cm) | ANCHORS | STONE SIZE (cm) | ANCHORS |
|---|---|---|---|
| **Marble** | | | |
| 76.2 X 19.1 X 2.5 | 2 | 188.0 X 19.1 X 2.5 | 4 |
| 76.2 X 50.8 X 2.5 | 4 | 188.0 X 50.8 X 2.5 | 8 |
| 76.2 X 101.6 X 2.5 | 6 | 188.0 X 101.6 X 2.5 | 12 |
| 152.4 X 19.1 X 2.5 | 4 | 213.4 X 19.1 X 2.5 | 5 |
| 152.4 X 50.8 X 2.5 | 8 | 213.4 X 50.8 X 2.5 | 10 |
| 152.4 X 101.6 X 2.5 | 12 | 213.4 X 101.6 X 2.5 | 15 |
| **Granite** | | | |
| 71.1 X 11.4 X 3.2 | 2 | 71.1 X 11.4 X 2.2 | 2 |
| 71.1 X 50.8 X 3.2 | 4 | 71.1 X 50.8 X 2.2 | 4 |
| 71.1 X 57.2 X 3.2 | 4 | 71.1 X 57.2 X 2.2 | 4 |
| 71.1 X 60.0 X 3.2 | 4 | 71.1 X 60.0 X 2.2 | 4 |
| 71.1 X 61.0 X 3.2 | 4 | 71.1 X 61.0 X 2.2 | 4 |
| 71.1 X 66.7 X 3.2 | 4 | 71.1 X 66.7 X 2.2 | 4 |
| 76.2 X 11.4 X 3.2 | 2 | 76.2 X 11.4 X 2.2 | 2 |
| 76.2 X 50.8 X 3.2 | 4 | 76.2 X 50.8 X 2.2 | 4 |
| 76.2 X 57.2 X 3.2 | 4 | 76.2 X 57.2 X 2.2 | 4 |
| 76.2 X 60.0 X 3.2 | 4 | 76.2 X 60.0 X 2.2 | 4 |
| 76.2 X 61.0 X 3.2 | 4 | 76.2 X 61.0 X 2.2 | 4 |
| 76.2 X 66.7 X 3.2 | 4 | 76.2 X 66.7 X 2.2 | 4 |
| 152.4 X 11.4 X 3.2 | 4 | 152.4 X 11.4 X 2.2 | 4 |
| 152.4 X 50.8 X 3.2 | 8 | 152.4 X 50.8 X 2.2 | 8 |
| 152.4 X 57.2 X 3.2 | 8 | 152.4 X 57.2 X 2.2 | 8 |
| 152.4 X 60.0 X 3.2 | 8 | 152.4 X 60.0 X 2.2 | 8 |
| 152.4 X 61.0 X 3.2 | 8 | 152.4 X 61.0 X 2.2 | 8 |
| 152.4 X 66.7 X 3.2 | 8 | 152.4 X 66.7 X 2.2 | 8 |
| 188.0 X 11.4 X 3.2 | 4 | 188.0 X 11.4 X 2.2 | 4 |
| 188.0 X 50.8 X 3.2 | 8 | 188.0 X 50.8 X 2.2 | 8 |
| 188.0 X 57.2 X 3.2 | 8 | 188.0 X 57.2 X 2.2 | 8 |
| 188.0 X 60.0 X 3.2 | 8 | 188.0 X 60.0 X 2.2 | 8 |
| 188.0 X 61.0 X 3.2 | 8 | 188.0 X 61.0 X 2.2 | 8 |
| 188.0 X 66.7 X 3.2 | 8 | 188.0 X 66.7 X 2.2 | 8 |
| 208.2 X 11.4 X 3.2 | 5 | 208.2 X 11.4 X 2.2 | 5 |
| 208.2 X 50.8 X 3.2 | 10 | 208.2 X 50.8 X 2.2 | 10 |
| 208.2 X 57.2 X 3.2 | 10 | 208.2 X 57.2 X 2.2 | 10 |
| 208.2 X 60.0 X 3.2 | 10 | 208.2 X 60.0 X 2.2 | 10 |
| 208.2 X 61.0 X 3.2 | 10 | 208.2 X 61.0 X 2.2 | 10 |
| 208.2 X 66.7 X 3.2 | 10 | 208.2 X 66.7 X 2.2 | 10 |
| 213.4 X 11.4 X 3.2 | 5 | 213.4 X 11.4 X 2.2 | 5 |
| 213.4 X 50.8 X 3.2 | 10 | 213.4 X 50.8 X 2.2 | 10 |
| 213.4 X 57.2 X 3.2 | 10 | 213.4 X 57.2 X 2.2 | 10 |
| 213.4 X 60.0 X 3.2 | 10 | 213.4 X 60.0 X 2.2 | 10 |
| 213.4 X 61.0 X 3.2 | 10 | 213.4 X 61.0 X 2.2 | 10 |
| 213.4 X 66.7 X 3.2 | 10 | 213.4 X 66.7 X 2.2 | 10 |

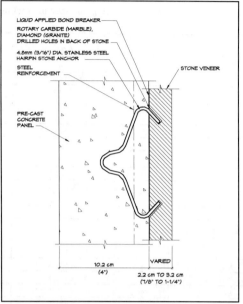

Figure 3. Hairpin Anchor

## FIELD INVESTIGATION

Due to reported stone cracking and displacement, and emergency removal of 25 loose pieces of marble, the authors were engaged to evaluate the stability of the facade, determine the causes of the observed behavior, design repairs to address deficiencies identified by the evaluation, and provide quality assurance monitoring of the repair construction. To accomplish these tasks, the first order of business was to determine what had caused the loosening of the 25 removed stones, and whether similar conditions were likely to be present elsewhere on the facade. The investigation focused on these pieces, the precast concrete panel locations from which they had been removed, and 30 additional stones randomly selected from marble and granite areas. Swing stages and telescopic boom lifts were used to provide the required access for detailed documentation and testing of the building facade, including core sampling of individual anchor locations for subsequent laboratory examination. A variety of defects were identified, including:

1.    Stone cracks were observed, including cracks running through the field of the stone anchors (leaving both pieces with independent anchorage to the precast concrete back-up), and corner cracks resulting in small unanchored stone pieces.

2.    Several previously removed stones exhibited poorly formed anchor holes at all anchor locations, with little or no ability to provide mechanical anchor engagement. Microscopic examination of the anchor holes detected cracking and spalling typical of installation with a percussion drill. Figure 4 shows the difference between properly and improperly drilled marble anchor holes.

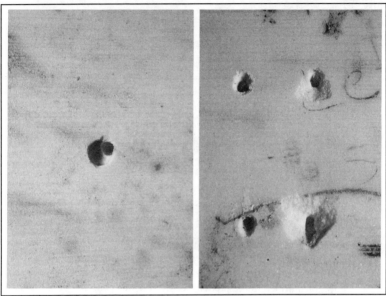

Figure 4. Photographs of Properly (left) and Improperly (right) Drilled Holes

3.    The anchors on several previously removed loose stones had disengaged from the precast concrete panel. Examination of the precast at these locations revealed these stones had apparently been replaced after the panels had been erected. Grout pockets had been chipped into the concrete, filled with a stiff grout mix, and the anchors on the replacement stone plunged into the plastic grout. Unfortunately, excessive grout stiffness precluded the grout from properly encasing the plunged-in anchors, leaving little mechanical attachment between the replacement stones and the precast concrete back-up.

4.    Additional stones beyond those previously removed were judged to be loose. These stones appeared to have been replaced after panel erection had been completed. Remedial anchorage of these stones to the precast concrete had been accomplished using 2.7 mm (7/64 inch) diameter mild steel wire ties fitted into edge holes at the four corners of the stones (refer to Figure 5). Many

of the wood wedges used to tighten the wires into the stone holes had weathered and fallen out, allowing the stones to rattle.

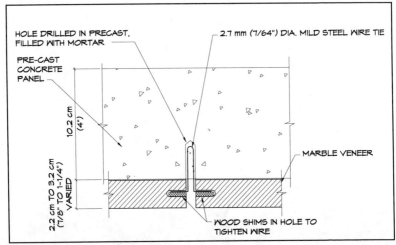

Figure 5. Wire Tie Remedial Anchor

5.    Another post-erection stone replacement method observed was the use of cadmium plated 0.6 cm (1/4 inch) diameter mild steel threaded rods. The rods were epoxied into the back sides of the replacement stones, and then grouted into holes drilled in the precast. Corrosion was observed on the unbonded portions of the rods.

6.    Honeycombing of the precast concrete in the vicinity of the stone anchors was observed at several locations (refer to Figure 6). In some cases the honeycombed area was extensive enough to significantly reduce, or negate, the capacity of the effected anchors.

7.    Twelve random core samples were taken to evaluate the drilled stone anchor holes on stones that had not been removed from the building. The cores were taken in such a way as to retrieve one anchor hole/anchor leg combination from a given anchor, without disturbing the rest of the connection. Curiously, one of the samples turned up a partially corroded stone anchor made of mild steel, instead of the specified stainless steel. The remaining 11 samples were identified as stainless.

Figure 6. Concrete Honeycombing at Anchor

## REPAIR CONCEPTS

The investigative findings caused concern both for the immediate stability of the cladding stones, and the long-term durability of the facade. Although the cracked stones could be readily identified for repair, the number and locations of the anchorage deficiencies were unknown. For any repair plan to be successful it would have to address all of the problems noted above, including the apparently "invisible" nature of the anchor flaws. Four basic remedial approaches were considered for the subject building:

1. Remove and replace the entire facade. This option would give the building a "clean slate" with regard to the stability concerns, and allow the owner the opportunity to affect aesthetic changes if desired, but at a very dear price.

2. Overclad the existing facade with another that is independently fastened to the building. This option offered advantages similar to facade removal and replacement, but at a lower cost.

3. Install auxiliary anchors of sufficient capacity to support each stone independent of the original anchors, and repair or replace the cracked stones. This option was cheaper than overcladding or replacement, but would have repaired many stones unnecessarily.

4.     Install auxiliary anchors only on those stones with deficient anchorage conditions, and repair or replace the cracked stones. This option potentially offered the lowest cost of the four approaches under consideration, but required a means of determining which stones were to be repaired.

The authors chose to pursue option 4 due to the large potential cost savings and the superior aesthetics generally associated with repairing the fewest stones possible. Identification of the cracked stones would be a straight forward matter, but the stone anchor deficiencies presented a more challenging problem since they did not lend themselves to detection by visual inspection or review of construction records. Means were necessary to nondestructively identify mild steel anchorage components, and to evaluate the in-place anchorage integrity of individual stones.

## NONDESTRUCTIVE ANCHORAGE MATERIAL IDENTIFICATION

The discovery of mild steel original and replacement stone anchorage components caused concern for the long-term durability of the facade. Unlike the mild steel reinforcing in the concrete areas of the cladding panels, the portions of the stone anchors within the stone do not enjoy passive corrosion protection from encasement in a high pH environment. Instead, moisture absorbed through the stone or leaking through failed stone-to-stone sealants can cause ongoing, and typically undetected, corrosion of unprotected mild steel. It is for this reason that stainless steel is the material typically specified for stone anchors, and the authors felt compelled to identify the locations of all mild steel stone anchors.

Normally the identification of metallic components within building construction is a reasonably straightforward task. Pachometers can be used to identify reinforcing bars encased in concrete, and other types of commercially available metal detectors can locate aluminum, copper, and stainless steel. By all outward appearances determining the locations of the mild steel stone anchors looked to be an easy task. Armed with an arsenal of half a dozen commercially available metal detectors, the authors quickly found this not to be the case. The pachometers that were tried, after tuning out the background "clutter" of the concrete reinforcement, simply were not sensitive enough to pick up the small legs of the stone anchors, which were "masked" by the heavier metal within the concrete back-up. The metal detectors that could tune out the background, and still remain sensitive enough to pick up the anchors, could not distinguish between mild steel and stainless steel.

To solve the anchor detection problem the authors resorted to developing their own equipment. After consulting with a number of instrument companies and component suppliers it was determined that a pair of inductive proximity sensors would probably work. One of the sensors was sized such that its maximum range was short enough to be unaffected by steel embedded in the precast concrete, sensitive enough to pick up mild steel hair pin anchors, wire ties, or threaded rods, but unaffected by standard stainless steel hair pin anchors. Fitted with a DC analog output circuit, this device, dubbed the "Analog Detector," had an output inversely proportional to the amount of mild steel within the first inch from the stone surface.

The second instrument developed had a detection threshold that could be adjusted to locate the individual legs of the stainless steel hair pin anchors. When the threshold was exceeded, the device indicated a positive read; otherwise no indication was present. The all-or-nothing nature of this instrument won it the name "Digital Detector." The performance and accuracy of both "home grown" instruments was verified by blind field trials conducted by GSA representatives. Mild steel and stainless steel anchor components were correctly identified despite the liberal use of background steel reinforcing bars and mesh in the test mock-up.

## IN-PLACE STONE ANCHORAGE EVALUATION

The defects uncovered during the field investigation suggested a high likelihood that insufficiently anchored, and possibly unsafe, stones still remained undetected on the building. The likelihood that groups of poorly formed stone anchor holes, honeycombed concrete at anchor locations, and inadequately anchored field replacement stones were present on the building caused concern for the stability of stones with these defects.

In order to pursue a repair addressing only those stones in need of remedial anchorage, a means of reliably locating these stones was needed. The first step taken toward this end was to determine the minimum number of anchors required to safely resist the wind and gravity loads applied to the cladding stones. Past experience by the authors indicated that most stone clad precast building panels are very conservatively designed when it comes to the number of anchors used on a given stone. If this trend held true for the building in question, it would mean that the cladding stones would still have adequate attachment to the building even if a number of anchors on a given stone were defective.

Although safety factors of 4.0 are common in stone cladding design, a minimum safety factor 2.0 was selected for this project, since the investigative intent was to verify the conditions on each and every stone. By carefully evaluating each stone, many of the variables a 4.0 safety factor is meant to address would be eliminated, allowing the proposed 2.0 minimum safety factor to cover the remaining unknowns. A detailed discussion of safety factors is beyond the scope of this paper, but is intended to be covered in a subsequent publication.

To determine the minimum number of "good" anchors required for each of the various cladding stone configurations on the building, an analytical approach using finite element computer analysis was chosen. The modeling techniques and laboratory testing performed to provide the required material properties and boundary condition parameters are discussed below.

**Laboratory Material Testing**

Computerized analysis of stone cladding panels requires accurate information regarding in-place component performance. Values were needed for stone flexural strength and modulus of elasticity, and anchor strength and stiffness. Laboratory tests were performed to measure these values for both the marble and granite on the project, with the results being incorporated into finite element analyses of variously sized stone cladding panels.

Stone flexural properties were determined by laboratory testing according to the methods of ASTM C880, *Standard Test Method for Flexural Strength of Dimension Stone* [1], and C1352, *Standard Test Method for Flexural Modulus of Elasticity of Dimension Stone* [2]. Test specimens were made from stone samples removed from the building. In order to check for anisotropic stone behavior, specimens were cut with their lengths in mutually perpendicular directions. Two-point loadings were applied at quarter points of all specimens as shown in Figure 7. Modulus values were derived from secant slopes on load versus deflection plots as required by the C1352 test method.

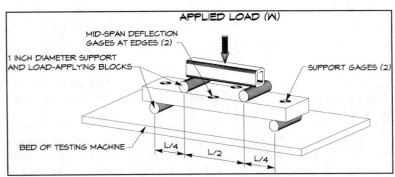

Figure 7. ASTM C-1352-96 Flexural Modulus of Elasticity Test

Anchor tension tests were used to define the tensile deformation behavior and ultimate capacity of the hairpin stone anchorages. The anchor test devised by the authors followed the general requirements of ASTM C1354, *Standard Test Method for Strength of Individual Anchorages in Dimension Stone* [3]. The testing apparatus was designed to apply incremental rotation-free displacements to the stone/anchor interface, thereby simulating in-place tensile loadings applied to stones with multiple hairpin style anchors. A custom-built steel frame (refer to Figure 8) fitted with two load cells and Teflon-treated guides was used to test anchorage specimens consisting of a 14 cm by 28 cm (5-1/2 inch by 11 inch) piece of stone with single centrally located hairpin anchor cast into a grout cube measuring 25 cm by 35.5 cm by 7.6 cm (10 inch by 14 inch by 3 inch).

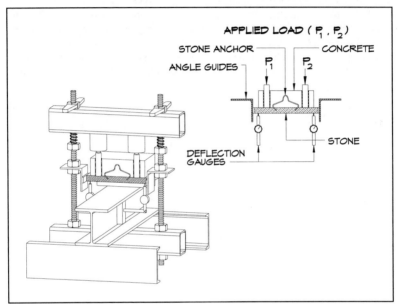

Figure 8. Stone Anchor Testing Apparatus

Load was applied to the anchor specimen through two solid steel rods penetrating the grout cube. Winged nuts were turned until a 0.25 mm (10 mil) displacement was measured on each side of the anchor, confirming a rotation-free separation of the stone from the grout block. The corresponding readings for each load cell were recorded at these 0.25 mm (10 mil) increments until the anchor specimen failed. Anchorage spring stiffnesses were based on plots of total load versus deflection.

**Critical Anchor Prediction**

Using the laboratory test results to provide member properties, a series of three-dimensional elastic finite element computer models were developed to evaluate the minimum number of "good" anchors required for each cladding stone configuration.

Based on the location and height of the building, three BOCA wind load zones were identified: 1.20 kPa, 2.39 kPa, and 2.87 kPa (25, 50, and 60 psf). Each stone size, type, and anchor configuration was modeled for the entire building, with the appropriate wind loads applied.

The overall geometry of the three-dimensional finite element computer model developed for one particular stone/anchor configuration is shown in Figure 9. Each computer model consisted of numerous element types; the precast concrete was modeled as rigid, with stiff axial compression only elements between the stone and concrete, the anchors were modeled using axial spring elements, and the stone was modeled using eight-noded plate elements with isotropic properties. The initial gap between the back side of the stone and the precast panel was modeled as 0.025 mm (0.001 inches), essentially zero, to simulate the typical conditions observed during field investigations.

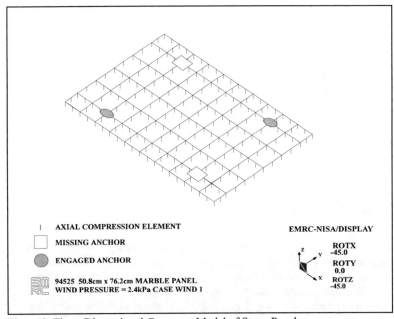

Figure 9. Three-Dimensional Computer Model of Stone Panel

Based on past experience [4], the computer model plate elements used the modulus of elasticity for the granite and marble panels based on an effective secant flexural modulus. The stone modulus was an average of the strong and weak direction stone flexural test results. The anchor elements were modeled with a spring stiffness which best simulated the actual anchor behavior detected during the anchor engagement laboratory testing. The representative model anchor spring stiffness was calculated using the load deflection curves generated during the anchor testing program. To simplify the task of analysis, it was desirable to model the anchorages as single elements with appropriate axial stiffness. Comparison computer models were generated with the stone anchors modeled as either one or two point boundary conditions, and the stone anchor stresses and deflections were compared. The results did not change significantly from one model to the other, therefore, the anchors were modeled using only one point to simplify the models. Table 2 summarizes the material property values used for the analyses.

### Table 2 - Finite Element Model Material Properties

| Material Property | SI Units | English Units |
| --- | --- | --- |
| Marble Flexural Strength | 8.1 MPa | 1,171 psi |
| Marble Flexural Modulus of Elasticity | 21.4 GPa | $3.1 \times 10^6$ psi |
| Granite Flexural Strength | 11.0 MPa | 1,590 psi |
| Granite Flexural Modulus of Elasticity | 29.0 GPa | $4.2 \times 10^6$ psi |
| Marble Anchorage Strength | 1.18 kN | 266 lb |
| Marble Anchorage Stiffness | 12.1 kN/cm | 6,900 lbs/in |
| Granite Anchorage Strength | 1.20 kN | 270 lb |
| Granite Anchorage Stiffness | 17.5 kN/cm | 10,000 lbs/in |

In order to evaluate a given stone/anchor configuration, the as-designed case was modeled first, followed by additional models with single or multiple anchors removed. In this way, the stone stresses and anchor loads for the original design and the "failed anchor" cases were both individually reviewed, and the interaction between anchor location and stone stresses was studied. The computer model results were compared with the laboratory-determined stone flexural strengths and anchor capacities. If the stone stresses and the anchor loads indicated by the analysis resulted in safety factors greater than 2.0, the panel configuration was accepted. Additional anchors were removed until either the stone stresses or the anchor loads exceeded the allowable limits. The anchors that remained were deemed "critical anchors." A minimum requirement of two anchors was also maintained. Figure 10 shows a representative stress plot for an anchor configuration which is missing two anchors at opposite corners.

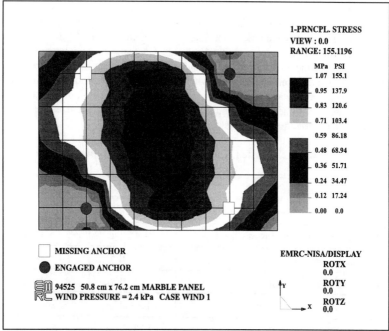

Figure 10. Stone Panel Principle Stress Plot

The critical anchor analyses verified the authors' prediction of excess capacity within the cladding stone system, particularly when safety factors of 2.0 were considered. In general, critical anchor locations were found to be at the corners of the stones and at the middles of the long sides, leaving a significant amount of redundancy in the system. At lower wind loads many of the stones only required engagement of two of the four corner anchors.

**Verification of Critical Anchor Presence**

The wind load critical anchor analyses revealed small but measurable changes in the stone panel deflections when individual anchorages were removed from the models. This behavioral change led the authors to consider stone deflection response to controlled loads as a possible means of identifying failed or ineffective anchor locations. To explore this idea further, a second series of computer models was generated. This time, however, each model considered all anchors fully effective except for a single critical anchor. Comparison of the stone deflections with and without the anchor effective again revealed consistent deflection changes large enough to be measured with precision dial gages, indicating instrumented field load testing could be used to detect failed anchors.

The methods of ASTM C1201, *Test Method for Structural Performance of Exterior Dimension Stone Cladding Systems by Uniform Static Air Pressure Difference* [5] are typically used as guidelines for performing in-place stone panel load tests. This test method, which utilizes a sealed chamber to apply simulated wind suction and pressure loads, was considered impractical for this project given the large number of stones involved (over 14,000). A simpler, more cost effective test was needed. A third series of computer models was generated to study the effects of a small, discrete tensile test load applied directly to a critical anchor location. Comparison models were created with engaged and missing anchors subjected to a 222 N (50 lb) tension load applied at the missing anchor location. The computer again indicated changes in deflection behavior large enough to detect with dial gage instrumentation. Comparison field testing on actual stone panels with known engaged and disengaged anchors was performed to verify the computer model accuracy. Table 3 summarizes the critical anchor deflections due to concentrated 222 N (50 lb) test loads.

**Table 3 - 222 N (50 lb) Critical Anchor Deflection Range**

| Condition | Anchor Effective | Anchor Ineffective |
|---|---|---|
| Marble Corner | 0.196 - 0.277 mm<br>0.0077 - 0.0109 in | 0.732 - 5.024 mm<br>0.0288 - 0.1978 in |
| Marble Mid-Side | 0.127 - 0.140 mm<br>0.0050 - 0.0055 in | 0.366 - 0.470 mm<br>0.0144 - 0.0185 in |
| Granite Corner | 0.132 - 0.229 mm<br>0.0052 - 0.0090 in | 0.432 - 6.363 mm<br>0.0170 - 0.2505 in |
| Granite Mid-Side | 0.074 - 0.135 mm<br>0.0029 - 0.0053 in | 0.196 - 0.587 mm<br>0.0077 - 0.0231 in |

## FIELD INSPECTION AND TESTING PROGRAM

Armed with the computer model stone deflection limits for the discrete 222 N (50 lb) load test and a proven procedure for stone anchor material identification, a field crew of four set out to identify the condition of each stone on the building. Customized recording sheets depicted the stone layout for one floor height by one bay width. Each sheet contained the location of the original stone anchors, and the planned tension load test locations. Each stone was tested for the presence of mild steel anchorages. Stone cracks, missing pieces, stone distress, and existing stone repairs were also identified and recorded. Finally, deflections due to the 222 N (50 lb) discrete tensile load tests were recorded for each critical anchor location.

The mild steel anchorage detection procedure utilized the two custom-built detection devices described in the *NonDestructive Anchorage Material Identification* section. The Analog Detector was first swept along the perimeter of each stone and then across the entire field of the stone. When mild steel was encountered, the observed voltage loss was recorded on the field data sheets. Next, the Digital Detector was used to determine the anchor configuration. If the Analog Detector sensed a mild steel anchor reading on a granite clad stone, the thickness of the stone was measured. When the stone thickness measured 2.2 cm (7/8 inch), a 0.3 cm (1/8 inch) thick plastic shim was attached to the analog sensor to simulate a stone of 2.5 cm (1 inch) thickness. If the anchor was not detected by the shim-equipped Analog Detector, the anchor was assumed to be a deeply-engaged stainless steel anchor. When mild steel indications were detected away from the typical anchor locations, the presence of each standard stone anchor was verified with the Digital Detector.

Stone load/deflection tests were performed on each exterior cladding stone at the predetermined critical anchor locations. As described in the previously, the load test locations were determined by finite element analysis for the appropriate design wind loads, providing a minimum factor of safety of 2.0 against stone or anchor failure. The field test procedure utilized a calibrated force meter attached to a suction cup to apply an outward force of 222 N (50 lb) at the test locations (refer to Figure 11). Deflections were measured with dial gages accurate to 0.025 mm (1/1000 inch). The dial gages were positioned with a suction cup on an adjacent stone. An electric vacuum pump supplied vacuum for the load application and dial gage suction cups. Dial gage locations, and primary and alternate load test locations for a specific sized stone are shown in Figure 12.

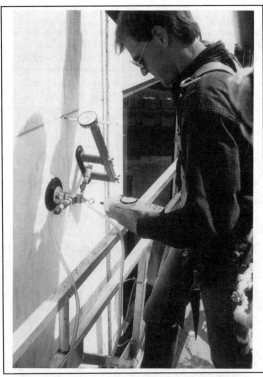

Figure 11. Field Crew Performing Tensile-Pull Test

The load application suction cup was placed at the primary test location, a 222 N (50 lb) load was administered, and the dial gage deflection was recorded. If the deflection exceeded the allowable limit, the nearest unused alternate location was tested. If the deflection at the alternate location exceeded the allowable limits, and no additional unused alternate locations existed, the stone was slated for auxiliary anchor installation. If a stone was cracked, each piece was considered an independent stone and load tested accordingly.

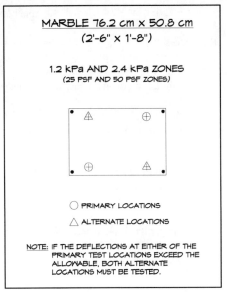

Figure 12. Load/Deflection Test Locations

## CONCLUSIONS

Each stone on the building was subjected to a visual inspection for cracking and previously repaired damage, an anchor material check, and load/deflection tests at each critical anchor location. Due to safety concerns expressed by the GSA, this work was performed during winter months and took five months to complete. Based on the inspection and test findings, auxiliary anchors were installed on approximately 600 of the 14,000 cladding stones. The total cost of the inspection and repairs was considerably less than any of the three alternate repair approaches discussed. Specific findings from this study are presented below.

1.      Nondestructive location and differentiation of mild and stainless steel stone anchorage components for stone clad precast concrete building panels can be performed using inductive proximity sensors.

2.      The stone anchorages on most precast building panels are conservatively designed with significant redundancy and excess capacity.

3.      The structural response of stone cladding attached to precast concrete building panels can be predicted by finite element analysis.

4.      Deflection response can be used as a nondestructive means of identifying failed or ineffective stone anchor locations on stone clad precast concrete building panels.

## Appendix

References:

[1]    ASTM C 880 - 96 Standard Test Method for Flexural Strength of Dimension Stone, American Society for Testing and Materials, Easton MD, 1996.

[2]    ASTM C 1352 - 96 Standard Test Method for Flexural Modulus of Elasticity of Dimension Stone, American Society for Testing and Materials, Easton MD, 1996.

[3]    ASTM C 1354 - 96 Standard Test Method for Strength of Individual Stone Anchorages in Dimension Stone, American Society for Testing and Materials, Easton MD, 1996.

[4]    Hoigard, K.R., Kritzler, R.W., Mulholland, G.R., "Structural Analysis of Stone Clad Precast Concrete Building Panels", International Journal of Rock Mechanics Mineral Sciences and Geomechanics Abstracts, Vol 30, No. 7

[5]    ASTM C 1201 - 96 Standard Test Method for Structural Performance of Dimension Stone Cladding Systems by Uniform Static Air Pressure Difference, American Society for Testing and Materials, Easton MD, 1996.

# Laboratory Evaluation of Building Stone Weathering

Seymour A. Bortz,[1] Bernhard Wonneberger[2]

Abstract

Dimension stone is among the most durable materials, but the process of weathering has shown that some types of stone, even of the same variety, are more durable than others. At present there is little information available about the durability of dimension stone on a building facade. Designers generally select a particular stone for its aesthetic qualities, with casual reference to basic parameters such as porosity, pore size, moisture absorption, and other critical physical and chemical parameters. Generally, when there is reference to weathering, the recommendation is to inspect another building with the same variety of stone. The recommendation does not consider the fact that stone, being a natural material, can vary considerably, even from one place in a quarry to another. Thus, in addition to observing weathering history in the field, we must determine how rock weathering can be recreated in the laboratory.

This paper attempts to provide background regarding the environmental processes that cause stone weathering in the field, such as acid rain (chemical), thermal (temperature changes), and freeze-thaw of absorbed water. We compare laboratory to field data that indicates accelerated durability testing can provide reliable information to long-term behavior of dimension stone.

## Introduction

Natural building stones are subjected to a variety of weathering conditions, both natural and man-made. Under these conditions, durability depends upon the stone's physical and chemical nature. Weathering of natural building stone consists of the reaction to environmental conditions within the body of the stone. It is the surface of

---

[1] Senior Consultant, Wiss, Janney, Elstner Associates., Inc., 330 Pfingsten Road., Northbrook, IL 60062-2095

[2] Senior Architect, Wiss, Janney, Elstner Associates., Inc., 330 Pfingsten Road., Northbrook, IL 60062-2095

the stone that is subjected to water and atmospheric gases. Weathering reactions are controlled by water and the natural gases dissolved in it (mainly $O_2$ and $CO_2$ and the human produced gases $SO_2$ and $NO$) that may penetrate the stone under various conditions. Weathering rates are influenced by temperature, moisture, organic acids, and dissolved carbon dioxide. The average rainfall is a major controlling factor of the weathering rate. The rain presents the dynamic for attack of the structure of the stone. This factor, in combination with the effects of our contemporary industrial society, accelerates natural weathering processes through elevated pollutant concentrations on the stone surface. Acid pollutants, in both air and rainfall, are recognized as serious hazards to carbonate rocks, such as limestone and marble, that are used in construction of major buildings. Silicate rocks and granite are affected by these acid pollutants to a lesser degree. However, even silicate rocks may be seriously affected by some acids and by alkalis. Elevated temperatures, or rapid temperature cycles, not necessarily cyclic freezing, affect differential volume changes of mineral grains. Temperature will accelerate solution of the carbonate minerals, while frost will cause damage to both carbonate and silicate stones.

Damage from frost is due to expansion of the freezing water. The structure of stone contains connected pores that can transfer water through the stone unit. Water entering the structure of the stone is an important consideration in this process. Rain, high humidity, and condensation are important factors in this weathering process. The water-filled pores exert forces against the pore walls by expanding water from ice crystallization during freezing. These pores may also contain clays, some of which may expand when wet. The swelling of expansive clay minerals when wet or by osmotic forces due to differential concentrations of dissolved material will put pressure against the pore walls. The outward forces of the expansion produce tensile stresses in the stone structure. Therefore, the tensile strength of the stone is of greater importance than the compressive strength with regard to freeze-thaw durability. For brittle material such as stone, the compressive strength is generally 10 to 15 times greater than the tensile strength.

## Variation of Dimension Stone Used for Building Facades

This section will give some background regarding the natural weathering of granite, marble, limestone, and sandstone. It is important to understand the differing characteristics of the various types of stone that are available. These characteristics affect the behavior of the stone material under natural weathering conditions.

Limestone is a rock of sedimentary origin, composed principally of calcium carbonate (calcite) or the double carbonate of calcium and magnesium (dolomite). Many limestones are formed of shells or altered shell fragments. Oolitic limestone, a popular building stone, consists of cemented rounded grains of calcite, generally less than 2 mm (5/64 in.) in diameter. Some limestones have varying amounts of other material, such as quartz sand or clay. These materials are mixed with the carbonate minerals. The carbonate materials will dissolve when exposed to acidic water or to water undersaturated with calcium.

Travertine is a variety of limestone usually precipitated from solution in ground waters and surface waters. When travertine occurs in hard, compact, extensive beds, it can be quarried and used as an attractive building stone. It is generally variegated gray, white, or buff, with irregularly-shaped pores distributed throughout.

Geologically, marble is a metamorphic rock formed by the recrystallization of the limestone or dolomite when subjected to high heat and pressure. Commercial marble includes all crystalline rocks composed predominantly of calcite and dolomite that can be polished to a reflective surface. It also includes serpentinite, which is not a carbonate rock. Thus, in addition to geological marble, commercial marble includes many crystalline limestones, travertine and serpentinite. In the metamorphic process, original sedimentary features, such as fossils, are usually destroyed, and the bedding planes are replaced by compositional layering (veins). The original calcite is recrystallized in an interlocking mosaic texture that provides the beauty a designer seeks for interior or exterior finishes.

Sandstone is a consolidated cemented sand sedimentary rock deposit. It has a distinctly granular texture with various cementing materials, including silica, iron oxides, and calcite. Enough voids generally remain in the rock to give it considerable permeability and porosity. (Commercially used sandstone is usually a sediment consisting almost entirely of quartz grains, 1 to 2 mm (1/32 to 5/64 in.) in diameter with various types of cementing material.) This allows the stone to readily absorb and confine water where it can freeze in colder weather.

Slate is a metamorphosed rock derived from argillaceous (clay) sediments consisting of extremely fine-grained quartz, mica, and other platy minerals. Slate possesses an excellent parallel cleavage that allows the rock to be split into thin, smooth-surfaced slabs with relative ease. The color of slate is generally determined by the oxidation state of pyrite (iron) or the presence of graphite. The cleavage planes are relatively weak in tension parallel to the planes. Water can enter and travel along these planes and later freeze in colder weather.

Commercial granite includes most rocks of igneous origin formed by solidification from molten rock beneath the earth's crust. True granites consist of feldspars and quartz, with varying amounts of other minerals such as mica and hornblende. These minerals are in an interlocking and granular texture with locked-in stresses. Each variety of mineral behaves differently when subjected to the same environmental conditions i.e. differential thermal stresses.

## Role of Chemical Weathering

Rainfall which enters the pores in the rock is a controlling factor of the weathering rate. Material can be dissolved by percolating water solutions or by the chemical decomposition of the stone. The amount of $CO_2$ in the water causes the aggressiveness of the water that can dissolve the stone. Weathering reactions also depend on temperature. Higher temperatures increase thermal agitation. Hence, such reactions are more magnified in the tropical areas than in colder temperature zones. In desert areas, the stone is hot but not subject to the intense action of liquid water weathering, so chemical weathering does not develop.

Table 1 shows the mean lifetime of 1 mm (1/32 in.) of unweathered stone when it is subjected to environmental weathering. These results show that in a cold, temperate, or tropical humid climate, the average rainfall (i.e., water rising) probably controls the rate of weathering.

Table 1.  Mean lifetime of 1 mm (1/32 in.) of unweathered stone [1]

| Stone Type | Climate | Lifetime |
|---|---|---|
| Acid (Light-colored igneous granite) | tropical semi-arid<br>tropical humid<br>temperate humid<br>cold humid | 65 to 200 years<br>20 to 70 years<br>41 to 250 years<br>35 years |
| Metamorphic (Marble, slate, etc.) | temperate humid | 33 years |
| Basic (Dark-colored igneous granite) | temperate humid<br>tropical humid | 68 years<br>40 years |
| Ultrabasic | tropical humid | 21 to 35 years |
| Sedimentary | arid-semiarid<br>all others | 50 to 100 years<br>highly variable |

Rates of silicate weathering are more difficult to evaluate because too many factors are involved which may influence the process. Stone exposed to polluted air may show signs of incipient weathering by undesirable discoloring, loss of polish and hardness of feldspars and ferro-magnesium silicates. There are estimates that there is a 2 to 2 ½ times increase of the weathering rate by solubilization and hydrolysis with each temperature increase of 10°C (50°F). The weathering rate can increase nearly 20 to 40 times in tropical moist areas.[2]

There have been observations that 10 mm (13/32 in.) of a limestone surface has been lost over a 300-year period of natural weathering, with about the same loss of a marble surface over a 150-year period. Sandstone with feldspar and mica as impurities will lose about 2.5 mm (5/64 in.) over 200 years, while almost pure quartz sandstone will remain sharp and clear over the same period.[3]

## Temperature Effects on Weathering

Weathering can also be caused by the difference of thermal expansion of the minerals that constitute the various stones. Volumetric or linear expansion of the different mineral materials that comprise the same stone mass can cause microcracking of the stone when heated in direct sunlight. Temperatures as high as 82°C to 88°C (180°F to 190°F) have been measured on dark stone surfaces. All minerals expand with increasing temperature, however, they expand to different degrees. Quartz expands about four times more than feldspars and twice as much as hornblende. Quartz is considered the most critical mineral under conditions of heating of granites and quartzitic sandstones. When quartz expands during heating, it exerts pressure against the surrounding crystals. These stresses can cause microcracking in the granite, lowering the strength and allowing other weathering phenomena, both chemical and freezing, to accelerate the disintegration processes. Figure 1 indicates the loss in compressive strength of granite when heated up to a temperature of 93°C (200°F).

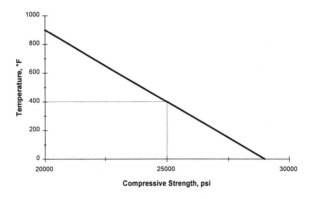

Figure 1. Loss in compressive strength of granite when heated.[4]

Individual stone crystals will have different expansion properties for each axis orientation, as shown in Figure 2. Calcite exhibits linear thermal expansion parallel to the C-axis of about 0.2 percent, but contracts about 0.1 percent perpendicular to the C-axis. The differing expansive rates of calcite cause microcracking in the structure of the stone. This disruption is manifested as increased volume and absorption of the stone during heating and cooling phases plus loss in strength. Stone that is low in quartz and carbonate minerals will expand very little with increases in temperature.

Another cause of stone disruption due to temperature changes is the differential expansion of trapped salts in the pores of the stone. Trapped salt can result from natural exposure due to high salt content in precipitation along coastal regions or from dissolved materials, (pollutants) in the environment. Figure 3 presents the thermal expansion of calcite, quartz, granite, and "rock" salt (halite). As the temperature increases from 0°C to 70°C (32° to 180°F), halite expands 0.5 percent compared with 0.2 percent for granite and about 0.1 percent for calcite. Small as these differences may appear, it is believed that expansion of absorbed salts combined with other factors can lead to stone decay.

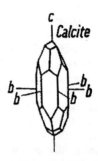

Figure 2.   Calcite crystal.

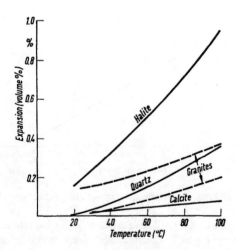

Figure 3.   Thermal expansion of calcite, quartz, granite, and "rock" salt (halite).[5]

## Effect of Freezing on Stone

Damage to stone results when temperatures are below freezing and any absorbed water forms ice crystals. The expansive crystallization pressure is very important as it produces tensile forces on the stone structure.

In addition to the expansion of the ice crystals, water itself expands just before it reaches a solid state. The ice volume is at a maximum and the density at a minimum of $4\,°C$ $(39\,°F)$ (from 1,000 kg/m$^3$ (62.4 lb/ft$^3$) in an unfrozen state to 916 kg/m$^3$ (57.2 lb/ft$^3$) in a frozen state). However, the density of water also increases with increasing outward pressure when it is in a confined space. This increase in density is similar to that of unconfined ice, but somewhat greater.

## Need for Laboratory Evaluation

The introductory sections of this paper were presented to provide a background for the design of a durability test procedure. This laboratory procedure has been used to develop information presented in the latter sections of this paper.

In the past, people have claimed that stone is "as hard as rock" and, therefore, durability testing is not necessary. However, as we have already shown, different types of stone will have varied behavioral characteristics when subjected to natural weathering. The long-term behavior can also be determined by using an accelerated test. The designer needs to know whether large blocks or thin slabs are to be used. Structural considerations for large blocks are mainly compressive strength. Because of the large mass of the stone units, minor loses of thickness and property changes due to weathering have a negligible effect on the structural capabilities of massive stone. However, structural considerations for thin stone include panel size, flexural strength, and weatherability. Loss of strength and small changes in thickness can have a major effect on the long-term behavior of thin stone facades.

A good laboratory test must consider environmental factors, such as temperature, air pollution, and rain. In addition, the designer must consider not only wind loads, but the fact that the stone can vary from quarry to quarry, from one area within a single quarry to another, and possibly within a single quarry block. Therefore, it is important to test the specific supply of stone for a large building project. Stone used successfully on a similar project in the past may not have the same physical or mechanical properties for the current project.

Based on previously discussed environmental effects on the properties of stone, we have developed a procedure that considers the following environmental factors: acid rain, temperature change, and freeze-thaw cycling. The test procedure consists of placing a stone specimen with minimum dimensions of 32 mm (1 1/4 in.) thick, 102 mm (4 in.) wide and 381 mm (15 in.) long in a 4 pH sulfurous acid solution. The specimens are immersed 6 mm (1/4 in.) to 10 mm (3/8 in.) deep in the solution in a stainless steel pan. Each specimen is also set on 6 mm (1/4 in.) diameter rollers to assure the stone face is subjected to the action of the bath solution. A fresh solution is used after each 25 cycle interval. The specimens are then subjected to 100 cycles between -23°C to +77°C (-10°F to +170°F). Before the test procedure is started, the test specimens are evaluated for dynamic Young's Modulus of Elasticity (sonic modulus) using ASTM Procedure C 215, "Test Method for Fundamental Transverse, Longitudinal and Torsional Frequencies of Concrete Specimens." The sonic modulus testing is repeated after every 25 freeze-thaw cycles to provide a nondestructive method of monitoring the changes in strength of the specimens. Figure 4 shows the relationship of sonic modulus to flexural strength developed from marble specimens. There is a good correlation at each measured cycle between strength and sonic modulus.

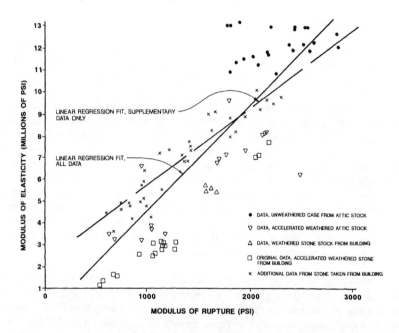

Figure 4. Relationship of sonic modulus to flexural strength developed
from marble specimens.[6]

## Laboratory Data

Table 2 provides stone durability test data obtained using the test procedure described above at the Wiss, Janney, Elstner Associates, Inc. (WJE) test laboratory over a period of ten years. Results from granite, marble, and limestone are presented. Note how stone types of the same classification vary, including stones of the same area and name. Most of the stones indicate strength changes during the durability testing. Table 3 shows how stones from different blocks of the same quarry can vary. Material from one quarry block withstands the duration of the test while material from another block does not. These results show the need for testing several different blocks of the same quarry for use on a single building project.

Table 2. Summary of Durability Test Results of Various Dimension Stones

| STONE | FINISH AND THICKNESS | INITIAL FLEXURAL STRENGTH kg/cm$^2$ (psi) | FLEXURAL STRENGTH AFTER TESTING kg/cm$^2$ (psi) | PERCENT LOSS OF STRENGTH |
|---|---|---|---|---|
| Spanish Pink Granite | 30 mm (1 3/16 in.) Thermal Finish | 94.6 wet (1,345) | Samples Fractured | 100% |
| Unidentified Granite | - | 115.3 (1,640) | 88.6 (1,260) | 23% |
| | | 115.3 (1,640) | 109.1 (1,552) | 5% |
| Moonlight Granite | 33 mm (1 5/16 in.) Thermal Finish | 74.5 wet (1,060) | 66.0 wet (939) | 11.4% |
| | | 80.5 dry (1,145) | 70.4 dry (1,001) | 12.6% |
| | | 74.5 wet (1,060) | 58.6 wet (834) | 21.3% |
| Stoney Creek Granite (Pink Granite from Bramford, CT) | 30 mm (1 3/16 in.) Light Thermal Finish | 79.4 wet (1,130) | 44.6 wet (635) | 43.8% |
| | | | 81.1 wet (1,154) | 17.3% |
| Mount Airy Granite (Fine-grained, white domestic granite) | 51 mm (2 in.) roughened finish | 98.4 wet (1,400) | 86.0 wet (1,223) | 12.3% |
| | | | 89.3 wet (1,270) | 9.0% |
| Venetian Gold Granite | 30 mm (1 3/16 in.) honed | 68.4 dry ‖ (973) | 54.1 dry ‖ (770) | 21% |
| | | 102.8 dry ⊥ (1,462) | 100.7 dry ⊥ (1,433) | 2% |
| Baltic Brown Granite | 30 mm (1 3/16 in.) honed | 95.4 dry ‖ (1,357) | 76.3 dry ‖ (1,086) | 20% |
| | | 112.2 dry ⊥ (1,596) | 98.9 dry ⊥ (1,407) | 12% |

Table 2. Summary of Durability Test Results of Various Dimension Stones
(Continued...)

| STONE | FINISH AND THICKNESS | INITIAL FLEXURAL STRENGTH kg/cm² (psi) | FLEXURAL STRENGTH AFTER TESTING kg/cm² (psi) | PERCENT LOSS OF STRENGTH |
|---|---|---|---|---|
| Rockville Beige Granite | 30 mm (1 3/16 in.) honed | 105.2 dry ∥ (1,496) 98.4 dry ⊥ (1,399) | a. 88.4 dry ∥ (1,257) b. 84.4 dry ∥ (1,201) a. 83.8 dry ⊥ (1,192) b. 76.6 dry ⊥ (1,089) | a. 16% b. 20% a. 15% b. 22% |
| Calibri Granite | 30 mm (1 3/16 in.) Polished Thermal | 128.3 dry ∥ (1,825) 190.3 dry ⊥ (2,707) 133.0 dry ∥ (1,891) 176.5 dry ⊥ (2,510) | 112.5 dry ∥ (1,600) 184.1 dry ⊥ (2,619) 136.8 dry ∥ (1,946) 173.1 dry ⊥ (2,462) | 12% 3% 3% gain 2% |
| Massangis Limestone | 51 mm (2 in.) | 112.0 dry (1,593) 79.0 dry ⊥ (1,124) 69.7 dry ∥ (992) | 95.5 dry (1,359) 71.1 dry ⊥ (1,012) 54.3 dry ∥ (773) | 15% 10% 22% |
| Chandore Limestone | 51 mm (2 in.) | 149.1 dry ⊥ (2,121) | 125.8 dry ⊥ (1,790) | 16% |
| Valdenot Limestone | 51 mm (2 in.) | 68.8 dry ⊥ (979) 81.0 dry ∥ (1,152) | 51.9 dry ⊥ (738) 83.7 dry ∥ (1,190) | 25% 3% gain |
| Luget Limestone | 51 mm (2 in.) | 65.9 dry ⊥ (938) 58.1 dry ∥ (826) | 0 dry ⊥ 0 dry ∥ | 100% 100% |
| Rosal Limestone | 51 mm (2 in.) | 81.5 dry ⊥ (1,159) | 0 dry ⊥ | 100% |
| Valders Buff Limestone | 51 mm (2 in.) | 109.0 dry ⊥ (1,551) 126.7 dry ∥ (1,802) | 122.8 dry ⊥ (1,746) 108.1 dry ∥ (1,537) | 13% gain 15% |
| Valders Dove White Limestone | 51 mm (2 in.) | 147.2 dry ⊥ (2,094) 125.4 dry ∥ (1,783) | 126.4 dry ⊥ (1,798) 100.2 dry ∥ (1,426) | 14% 20% |
| Valders Gray Limestone | 51 mm (2 in.) | 178.2 dry ⊥ (2,535) 147.6 dry ∥ (2,100) | 162.8 dry ⊥ (2,315) 115.9 dry ∥ (1,648) | 9% 22% |

Table 2.  Summary of Durability Test Results of Various Dimension Stones (Continued...)

| STONE | FINISH AND THICKNESS | INITIAL FLEXURAL STRENGTH kg/cm² (psi) | FLEXURAL STRENGTH AFTER TESTING kg/cm² (psi) | PERCENT LOSS OF STRENGTH |
|---|---|---|---|---|
| Dolomitic Limestone from France | 32 mm (1 1/4 in.) | 156.3 (2,223) | 120.9 (1,719) | 23% |
|  |  | 61.2 (870) | 38.8 (552) | 37% |
| Indiana Limestone | 64 mm (2 ½ in.) | 65.6 (933) | 59.3 (843) | 9.6% |
| Valders Dolomitic Limestone | 30 mm (1 3/16 in.) | 148.0 (2,105) | 153.3 (2,180) | 4% gain |
| Pierre De Lens Limestone | 51 mm (2 in.) | 89.5 dry ⊥ (1,273) | 68.1 dry ⊥ (969) | 24% |
|  |  | 59.3 dry ∥ (843) | 45.5 dry ∥ (647) | 23% |
| White Marble from Carrara, Italy after 15 years of exterior exposure | 32 mm (1 1/4 in.) polished | 64.0 wet (911) | 38.0 wet (540) | 41% |
|  |  |  | 41.4 wet (589) | 35% |
|  |  |  | 59.7 wet (849) | 7% |
| Georgia Golden Vein Marble | 25 mm (1 in.) honed | 79.0 dry ∥ (1,124) | a. 74.6 dry ∥ (1,061) b. 76.4 dry ∥ (1,087) | a. 6% b. 3% |
|  |  | 104.3 dry ⊥ (1,484) | a. 96.6 dry ⊥ (1,374) b. 97.3 dry ⊥ (1,384) | a. 7% b. 7% |
| Marquis Gray Danby Vermont Marble | 30 mm (1 3/16 in.) honed | 60.5 dry ∥ (861) | 54.1 dry ∥ (769) | 11% |
|  |  | 99.5 dry ⊥ (1,415) | 46.1 dry ⊥ (656) | 5% |
| White Carrara Marble | 25 mm (1 in.) honed | 154.4 dry (2,196) | 112.0 dry (1,593) | 27% |
|  |  |  | 90.6 dry (1,289) | 41% |

∥      Tested parallel to the bedding plane or rift
⊥      Tested perpendicular to the bedding plane or rift

Table 3.  Summary of Durability Test Results for Limestone
Removed from the Same Quarry

| Block A Tested Perpendicular to Bedding Plane | | |
|---|---|---|
| Cycles | 0 | 100 | Change |
| Dry | 81.5 kg/cm$^2$ | Unable to test | -100% |
| Wet | 43.2 kg/cm$^2$ | Unable to test | -100% |
| Block B Tested Perpendicular to Bedding Plane | | |
| Dry | 178.2 kg/cm$^2$ | 162.8 kg/cm$^2$ | -9% |
| Wet | 172.4 kg/cm$^2$ | 147.5 kg/cm$^2$ | -14% |

Figure 5 shows the change in sonic modulus that occurred when granite was subjected to the 4 pH sulfurous acid bath. Exposure to acid bath cycles apparently has little or no effect on the granite. However, the temperature change does appear to cause differential expansion that breaks the bond between the mineral crystals and, by that, lowers the strength. Figure 6 is a similar curve for marble. The change in property is due to differential expansion and contraction of the individual calcite crystals, and some dissolving of the calcite. Although some calcite dissolves in the acid solution, this essentially neutralizes the acid. Figure 7 consists of curves for different limestones. The Massangi and Valders dolomitic limestones show no basic effect while the Indiana Limestone has a slight downturn at the end of 100 cycles indicating dissolving of the calcite.

**Accelerated Durability Data**
**Rockville Beige Granite**

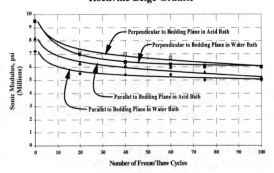

Figure 5.  Accelerated durability test results for
Rockville Beige Granite.

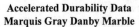

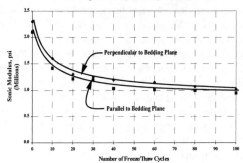

Figure 6.  Accelerated durability test results for
Marquis Gray Danby Marble.

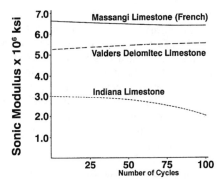

Figure 7.  Accelerated durability test results for three
different types of limestone.

## Natural Weathering Studies

The durability test allows one to make comparisons between stones.  However, there has been criticism that the durability test procedure has no relationship to natural weathering.  Arguments also state that the test has no meaning for building projects that are in warmer climates.  To counteract these comments, we have compared sonic modulus test results from stone subjected to natural weathering to sonic modulus test results determined from stone subjected to the durability test procedure.  For warmer climates, the test procedure can be modified to cycle between +5°C and +77°C (+41°F and +170°F).

Thirty-five years ago, twelve domestic marbles were placed on the roof of a building located immediately south of the main business district in Chicago. The marbles were monitored quarterly over 8 years using sonic modulus testing. Figure 8 is a chart showing the results of this work, including the results for.Danby marble, indicated as "I" on the chart. Recently, we had an opportunity to perform durability tests on a second set of Danby marble. The previously shown Figure 6 presents the change in sonic modulus determined during these tests. Figure 9 presents the natural weathering and durability test curves shown in Figs. 6 and 8. These curves show 100 freeze-thaw cycles of durability testing can be considered equivalent to 6 or 8 years of natural weathering. Therefore, 12 to 16 freeze-thaw cycles would be equivalent to one year of natural weathering in a northern temperature environment. Our data from this work and similar additional work using naturally weathered stone from a building was compared with data obtained from durability testing of attic stock stone, (stone kept in reserve, but not exposed to weathering). This work indicated that real-time effects could be estimated from laboratory tests.

Field Studies

Evaluation of a 10-year-old marble-clad high-rise office building in Rochester, New York, permitted evaluation of actual strength degradation from natural weathering. Many panels from the building facade, and panels that had not been exposed to natural weathering, were provided for durability tests. This provided a large statistical population for data analysis. The data is summarized in the chart previously shown in Figure 4. The chart shows a correlation between modulus of elasticity and flexural strength. A weathering chart was plotted using the modulus of elasticity and flexural strength data, Figure 10. In this figure, the flexural strength chart has been superimposed over the elastic modulus chart.

The process of obtaining the curves shown is empirical. To provide guidance regarding the meaning of accelerated weathering, the data from weathered and unweathered stone had to be related in some manner. Data was plotted for elastic modulus vs. accelerated aging cycles for both the unweathered and weathered stone. They were analyzed such that the original elastic modulus and strength for the building weathered stone are close to being a tangent to the curve for elastic modulus and strength of attic stock subjected to the durability test. This occurred at approximately 160 cycles. Using these relationships, the conclusion was drawn that 160 cycles of accelerated weathering appear to be equivalent to 10 years of natural weathering on the building. Thus, 16 cycles were determined to be equivalent to one year of service life in upper state New York. While this cannot be considered a rigorous proof of the time relation of natural aging to accelerated aging, the empirical data from the laboratory and field observations indicate this is a reasonable approximation of what the designer can expect regarding changes in strength properties from natural weathering.

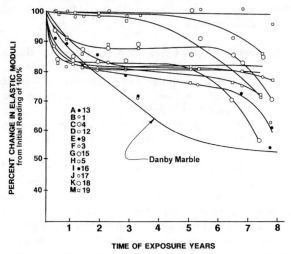

Figure 8. Chart showing natural weathering test results for twelve different domestic marbles.

### Estimated Life of Marquis Gray Danby Marble

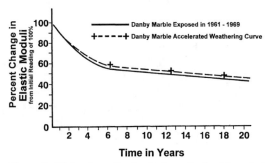

Figure 9. Natural weathering test results for Marquis Gray Danby Marble.

Using this technique, two other marble-clad buildings were also studied. A correlation between elastic modulus and strength was also determined, Figure 11. Composite curves for building weathered and attic stock were developed, Figs. 12 and 13. For the marble in Figure 12, 12 ½ cycles of accelerated weathering were found equivalent to one year of natural weathering on a building in Kansas City. For the marble in Figure 13, 13 cycles of accelerated weathering were seen to be equivalent to one year of natural weathering. The plot in Figure 13 is based on actual flexural strength of the stone, not sonic modulus. Note the curves for sonic modulus are similar to the curves for flexural strength.

The three marbles tested indicate approximately 15 cycles of accelerated weathering is equal to one year of natural weathering on the buildings. All the data correlates with the rooftop test and laboratory accelerated weathering of the Danby marble previously discussed. The Danby data showed that changes in properties due to natural weathering and accelerated weathering are similar. Curves from testing stone from the buildings and laboratory accelerated weathering can be tied together so as to predict the number of cycles that will constitute one year of weathering.

Limited data has also been obtained for granite and limestone. The results of these tests have been similar to those obtained from the more extensive marble data. Figure 14 presents a summary of all the data we have for granite, marble, and limestone. In this figure, changes in properties of granite were plotted for 300 cycles, 500 cycles for marble and 200 cycles for limestone. Based on the data, approximately 13 cycles represent one year of natural weathering for granite, 12 ½ cycles represent one year for marble, and 12 cycles represent one year for limestone.

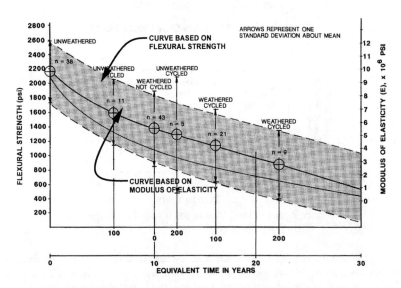

Figure 10.  Relation of number of durability test cycles to years
of natural weathering for Bianca White Carrara marble.
n=number of specimens.

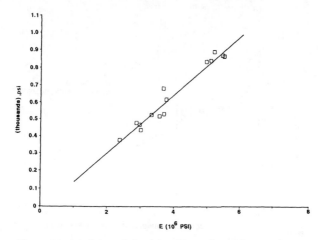

Figure 11.  Modulus of elasticity (E) vs. flexural strength
determined from marble specimens.

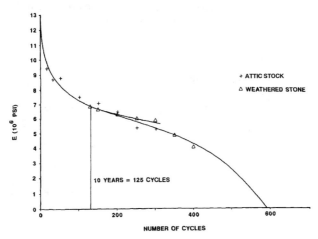

Figure 12.  Number of durability test cycles vs. modulus of elasticity (E)
determined from Georgia Golden Vein marble specimens.

We are currently exposing granite, marble and limestone specimens to natural
weathering on the roof of a WJE building in Northbrook, Illinois, Figure 15.  Figure 16
shows the sonic modulus curves determined after 1 ½ years of exposure.  The curves can
be compared with the data plotted in Figs. 5 through 7 for accelerated weathering.  Note
the similarity of the curves obtained during earlier accelerated weathering and the
naturally weathered stone.  These tests will be extended over a ten-year period.

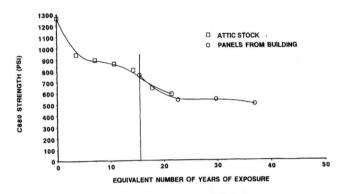

Figure 13. Comparison of durability test cycles to years
of natural weathering determined from White
Carrara marble specimens.

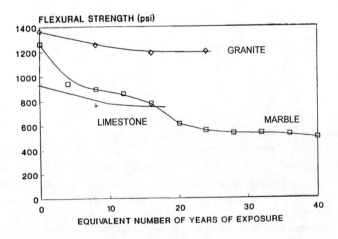

Figure 14. Flexural strength vs. years of natural weathering
for marble, granite, and limestone.

Figure 15. Stone test specimens exposed to natural weathering on roof.

Conclusions

    The efforts described in this paper include the results from several years of durability (accelerated weathering) testing. These tests were performed to determine the long-term durability of thin stone to natural weathering. The accelerated weathering test procedure described in this paper will distinguish between a durable and less durable stone under natural weathering conditions. The procedure will also provide an indication of strength loss due to weathering. This data can affect the structural design for long-term reliability of thin stone panels on high-rise buildings.

    The data obtained showed useful information can be obtained from thin stone specimens less than 50 mm (2 in.) subjected to the described accelerated aging test for at least 100 cycles. If comparison tests can be made using naturally weathered stone and unweathered stone, a relationship can be determined between number of cycles and time. The design can then be optimized by either increasing the thickness of the panel or reducing the unsupported span. These options will reduce load stresses and increase the service life of the stone.

**Natural Weathering Study Test Results**

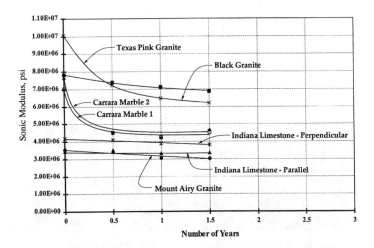

Figure 16.  Number of years of natural weathering vs. modulus of
elasticity (E) determined from various stone specimens on roof.

The weathering curves provided can be used for design by determining where
the allowable working stress crosses the strength loss curve for the stone. When data
for naturally weathered stone is not available, an estimated cycle per year can be used
from the data in this paper. Assume 12 to 15 freeze-thaw cycles of the durability test
procedure is equivalent to one year of natural weathering in a temperate climate.

We would like to acknowledge Mr. William G. Hime, Senior Consultant with
Erlin, Hime Associates, (EHA) and Mr. Ross A. Martinek, Senior Petrographer with
EHA, for their helpful review and comments in the preparation of this paper.

References

(1)  Nahon, Daniel B.  *Introduction to the Petrology of Soils and Chemical
Weathering.*  New York:  John Wiley & Sons, Inc., 1991.
(2)  Ibid., page 6.
(3)  Winkler, E.M. "Important Agents of Weathering for Building and Monumental
Stone," *Engineering Geology I*, 1966, pages 381-400.
(4)  Winkler, E. M. *Stone: Properties, Durability in Man's Environment.*  Second,
revised edition, Wien, New York:  Springer-Verlag, 1975.
(5)  Ibid., page 125.
(6)  WJE in-house testing.

# Edge Spalling In Thin Stone Cladding

## James H. Larkin[1]

## Abstract

Current trends toward the use of thinner and thinner natural stone as an exterior cladding material warrant new design considerations and demand higher levels of quality control and possibly new methods of fabrication.

A combination of the desires to reduce weight and cut budgets has driven the stone cladding industry into new territory in the last twenty years. It is now common for stones to be slabbed to a 3cm thickness, and in some cases to 2.5cm and less. In addition to becoming inherently more fragile, the reduced thickness leaves very little room for error in edge preparations. Kerf cuts and dowel pin holes must be fabricated with very tight precision. The author and his firm have observed dozens of stone-clad buildings in the last ten years with localized damage or spalling at the edges of individual pieces of stone. The spalls are typically semi-circular in shape and centered on the stone's anchors back to the structure. They have been observed in granite, marble and travertine (Figures 1 & 2) and have been observed domestically as well as on buildings in Southeast Asia.

In this paper, causes of this phenomenon are explored and the effects on the veneer's useful life are discussed. Several case studies are referenced in order to draw the following conclusion:

While some stone damage is service oriented, i.e., occurs during the life of the cladding after erection, it appears evident that a majority is related to anchor design, fabrication, handling, and/or installation procedures. Design emphasis for these thin veneers must shift from strictly analyzing the stone's free span to more closely considering the concentrated reactions at their support points.

[1]Senior Design Consultant, Curtain Wall Design and Consulting, 9101 LBJ Freeway, Suite 500, Dallas, Texas 75243

The paper discusses such design considerations. The prevalence of spalling at support points indicates that the industry has, without improved design or quality control, reached or possibly passed the point at which we can continue to reduce stone thickness and successfully install structurally adequate cladding systems.

## Introduction

Exterior cladding engineers are responsible for determining appropriate stone thicknesses. The bulk of their design effort is typically spent on checking mid-span thicknesses against design wind loads for vertically oriented stone. In some cases the thickness will be increased to account for aging and a loss of strength due to exposure and the effects of weathering (Widhalm 1996). With their concentration on flexural design, however, engineers are overlooking some important local effects at the stone's connection points.

If flexural strength were indeed the governing design factor, one would expect stone failures or breakage to occur mid-span or between support points. This has not been this author's observation.

As an owner's consultant, the author's firm has observed buildings with broken stone domestically in Texas, Tennessee, Georgia, North Carolina, Florida, Illinois, and Washington DC, and internationally in Singapore and Hong Kong. In most all cases, the root source of failures seems to be traceable to poor anchor design and/or fabrication/installation procedures.

The objective of this paper is to present the case that design emphasis should shift to edge conditions of thin stone veneers as opposed to solely analyzing mid-span flexural strength under design load (Bayer 1993). Perhaps existing projects have adequate safety factor or have yet to be subjected to design load events, but mid-span failures have not surfaced as the major concern for stone as a cladding material (Clift 1989). Localized spalling or cracking in the vicinity of edge supports on the other hand is a real problem.

## Observed Performance of Thin Stone Cladding

One prominent, travertine clad, tower investigated by the author did survive a wind event which approached or exceeded its stone's design forces. The 3cm travertine had been exposed to 10 years of the salt air and high UV environment of downtown Miami, Florida when Hurricane Andrew struck in 1992. The eye of the storm crossed land about 15 miles south of downtown, but there was significant damage to heavy equipment on the roof of this particular building to evidence dangerous wind forces.

Throughout the storm, the travertine cladding performed relatively well. Less than two dozen out of 12,000 stones were identified with damage attributable to Andrew. Some of this damage was directly related to impact from flying debris and other to errors by the construction crews during the original erection (stone anchors were mislocated). In general, the system performed relatively well and one could not conclude that insufficient stone thickness was used based on flexural strength design for the material.

Conversely, many stone clad buildings are being observed with edge spalling. The concerns with this type of stone damage are 1) that the spalls are small and difficult to detect visually on tall towers, 2) that if spalls exist on the exposed face of the stone, there is likely more unseen damage to the inboard face, and 3) that spalls are located near the anchor points for individual pieces of stone. Damaged stone on the inboard face or "backside" located near an anchor is the most dangerous. As perimeter caulking ages and as wind and other structural forces work on these weakened areas of stone, the risk that a piece could fall from a building increases.

## Re-directed Design Emphasis

In order to reduce the incidence of localized edge damage, cladding engineers should consider the four contributing factors identified and discussed below. Specifically the degree of localized edge damage is a function of:

1.   Anchor Design
2.   Quality Control in Stone Fabrication
3.   Care in Shipping and Handling
4.   Service Life Conditions

All of these factors have become increasingly more important with the use of 2 and 3 cm thick stones as an exterior cladding material. First of all, the type of anchor or its design effects the performance of the stone it supports.

Whether dowel holes or kerf cuts can safely be fabricated into the thin stone edge, or whether backside support is necessary, is of prime importance to the designer. All other factors effecting this localized damage are a function of the anchor design. Probably the next most important consideration is Quality Control. Once the design has been selected, precise fabrication is essential to avoid stone breakage. The required level of precision in stone fabrication increases as thickness decreases. Further risk of damage is then incurred as the fabricated stones are transported to the job-site and erected in place. Finally, there are service life conditions that determine the degree of damage a piece of stone may or may not experience.

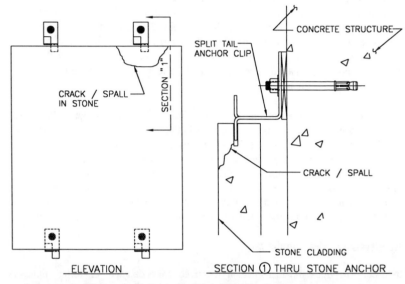

**Figure 1.** Schematic view of localized edge damage.

**Figure 2.** The edge spalling of the stone in this photograph is a common problem on many thin stone cladding systems.

## 1. Anchor Design

As mentioned above, the type of support mechanism the designer chooses effects the performance of his stone cladding. A preferred method would employ the use of kerf clips of 100 to 150 mm in length minimum, as opposed to dowel pins or split tail anchors (Marble Institute of America 1991) that engage a 20 mm wide tab into the edge of the stone. This preference is based on the capacity of the clip and its ability to distribute load over a larger area. The author has observed a higher incidence of edge damage on cladding systems secured with dowel pins and narrow tabs than on those with longer kerf clips. Concentrated point loads at dowel pins or narrow tabs should invoke the use of higher safety factors by the engineer (Clift 1993).

The design of the anchor can also effect the erectability of the stone. Many of the buildings observed by the author with edge spalling rely on adjacent stones sharing a common dowel pin. The problem with this design begins with the stone fabrication and manifests itself with the installer forcing one stone down onto another when their dowel pin holes do not align.

Figure 3 shows how dowel pin holes are laid out relative to one end of a piece of stone. Note that it is possible for stones, which are cut and drilled within acceptable tolerances, to have dowel holes misalign because of the tolerance in the stone's overall length. Holes that share a common pin could fall 372 mm from the end of stone on one piece and then 378 mm from the end of stone on the adjacent piece. This example is shown Figure 4.

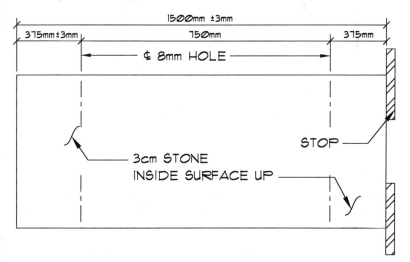

**Figure 3.** Plan view of a stone slab dimensioned for anchor hole fabrication.

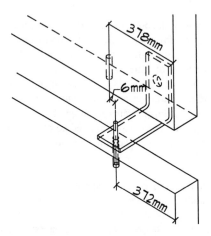

**Figure 4.** Stones sharing a common dowel pin can misalign due to fabrication procedures.

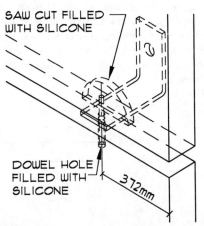

**Figure 5.** A groove or saw cut in one edge can accommodate the misalignment as noted in Figure 4.

Figure 5 identifies a modification to the common dowel pin design which has been successfully used to avoid this misalignment problem. Without the flexibility offered by the groove or saw cut in the upper stone, stone setters will use prybars to bend pins and otherwise force the two stones together. This prying by the stone setter is a major cause of the edge spalling problem and a problem that the designer needs to address.

A common, but undesirable, solution to the misalignment problem with dowel pin supports is to field-drill the holes in the edge of the stone. The leads to the second mentioned contributing factor for localized edge spalling, "Quality Control".

## 2. Quality Control in Stone Fabrication

It is imperative that all stone anchors be fabricated under controlled factory conditions. Holes should be drilled with liquid cooled, diamond coated, bits as opposed to impact hammers. The drilling must be mechanically guided and not done with hand held tools. There is too little room for error, when stone thicknesses approach 3cm and less, to rely on field fabrication and hand drilling.

The contractor on the 25-story, Ritz-Carlton Hong Kong hotel (completed in December of 1991) ran into the misalignment problem mentioned above, early in his stone installation. He quickly shifted to field drilling 8mm diameter holes into the edge of his 25mm thick granite. In so doing, he lost his ability to control the quality of drilling. Even small amounts of tolerance in hole location and alignment relative to the face of the stone were significant in this 25mm stone.

As a result of slightly skewed holes (see Figure 6), pins pried on the thin stone edges when they were set into place and over 1½% of the buildings 20,000 stones were observed to be broken. The contractor used epoxy to repair visible damage, however the owner was left with a new building that was speckled with unsightly patches and a veneer of questionable durability.

More importantly, there are surely spalls on the inboard surface of many stones that have gone undetected and unrepaired. This backside damage is the most significant concern. With the loss of sound stone behind the dowel pin, suction forces on the facade could lead to a piece of stone falling from the wall!

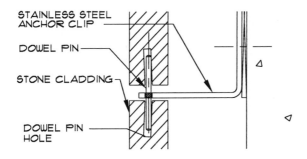

**Figure 6.** Poorly drilled dowel pin holes.

The Austin Center Building in Austin, Texas suffered similar problems as the Hong Kong hotel, with a different type of stone anchor. In this case, granite was hand set to a concrete back-up with disk anchors engaging a semi-circular groove cut into the edges of the stone.

The original stone thickness was value-engineered down to 20mm and hand held saws were used to field-cut grooves into its edges to accept the disk anchors. Both of these factors ultimately led to wide spread spalling and thus stone replacement soon after the building was completed.

Figure 7 illustrates what was observed on some of the stones removed from the wall at Austin Center. The cuts varied in width (more than one half of the stone's thickness was removed by the saw cut in some cases) and alignment. Again, thin stone veneer necessitates extremely tight Quality Control that is difficult to achieve in the field.

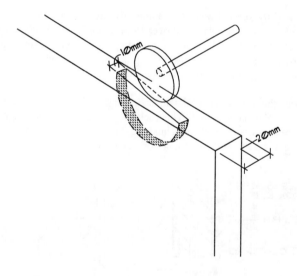

**Figure 7.** Saw cut is not centered and parallel to face of stone.

## 3. Care in Shipping and Handling

Good designs and well fabricated stones are still subject to impact and other handling loads that can cause breakage prior to being final set onto a building. Some of the most severe load conditions that stone, supported by steel trusses, ever experiences occurs prior to installation. The 26-story limestone building erected on the Chicago campus of Loyola University in 1993 lost 15% of its 50mm thick limestone to damage incurred during shipping and handling.

Precast concrete panels must be designed for lifting loads when they are removed from their form-work. Generally, if they arrive onto the building uncracked, they are then over-designed for any service loads they will ever experience. Thin stone that is set onto steel truss work in the factory is very similar. Stone damage in this type of design is generally a result of vertically oriented panels (column covers) being hoisted and moved horizontally, or without spreader bars, or because of impact loads during shipment, or other similar effects.

The engineer must design frames that can handle the twisting and eccentric load conditions that will be experienced during transit from the fabrication shop to the job site. Additionally, stone anchors must be resilient or flexible enough to accommodate or absorb the related movements and shock.

## 4. Service Life Conditions

The majority of stone damage observed by this author is localized at edge supports, even that which is attributable to forces on the stone after it is in service. Most of the damage in this category relates to insufficient consideration given to structural movements. Spalling of the cladding can occur when the support structure moves under wind loads, during seismic events, or in the case of concrete buildings, when the structure shrinks. Similar to the previously mentioned consideration for accommodating movements during shipping and handling, stone anchors must allow for, or absorb, the structure's movement.

It is important to use an elastomeric material to cushion the interface and absorb relative movements between metallic anchors and brittle stone. Where epoxy is used, within dowel pins or kerf cuts, prying forces from the structural movements can pry and break the stone. The cladding on the Ritz-Carlton Hong Kong mentioned above will likely experience further damage from normal shrinkage of its concrete structure. Creep and column shortening, which will continue with time, will impose these prying forces onto the stone.

The pins are epoxied into their holes and will not allow the caulk joints between stones to compress as the structure shortens. Instead, the stainless steel clips bolted to the structure (see Figure 5) will rotate with respect to the plane of the stone and cause the dowel pin to pry on the edge of the stone. Cladding designers must anticipate these structural movements and choose soft supports for their brittle stone.

## Summary

Flexural strength is a critical design parameter for sizing stone as an exterior cladding material. It is all too often the only consideration that a cladding engineer will use in his analysis. Some engineers will expand their analysis in an attempt to account for the effects of aging of the stone. However, one of the main causes of failures in thin stone cladding systems is routinely overlooked. Anchor designs and stone edge conditions have become very important considerations for today's cladding engineer.

Even with the current trend toward very thin veneers, the author is aware of very few instances of stone failure which have resulted from excessive flexural stress. Edge spalls or localized breaks at the stone anchors on the other hand are a prevalent condition on many buildings world wide. They are not easily detected on the outside of tall buildings and are almost impossible to identify on the backside of the cladding. The localized damage does exist however and there is a need for cladding engineers to shift their emphasis to minimize the risk of stones falling off of a building.

## Keywords

Stone, exterior cladding, edge spalling, flexural strength, dowel pin, kerf.

## References

Allen, D. and Bomberg, M. (1996). *Use of generalized limit states method for design of building envelopes for durability.* Thermal Insul. and Bldg. Envs. Vol. 20, July 1996. 18-39.

Bayer, J. and Clift, C. (1993). *Use of physical property data and testing in the design of curtain wall.* Int. J. Rock Mech. Min. Sci. & Geomech. Abstr. Vol. 30, No. 7, 1563-1566.

Clift, C. and Bayer, J. (1989). *Stone safety factors: much ado about nothing.* Dimensional Stone, January/February 1989. 38-40.

Widhalm, C., Tschegg, E., Eppensteiner, W., (1996). "Anisotropic thermal expansion causes deformation of marble claddings." *Journal of performance of constructed facilities*, ASCE, Vol. 10, No. 1, 5-10.

Marble Institute of America (1991). *Dimensional Stone Design Manual IV.*

# Assuring the Durability of Stone Facades in New Construction

by Michael D. Lewis, AIA[1]
and William H. McDonald[2]

## Abstract

The modern building often has a facade that performs poorly. Too many facades built during the last thirty years, and intended to be weathertight and attractive for generations, now need major structural rehabilitation just to make them safe. Their leaks, drafts, and dilapidated states are accepted only because these problems are common. Ironically, hundred year old stone facades are out-performing many that are one-tenth their age.

Problems occur in modern facades because their designers rarely understand their complex function or know how they're built. The exterior wall evolved away from solid stone into a stone cladding on backup when the facade was fully separated from structure fifty years ago. Weather tightness was considered less important than exploiting aesthetic design freedom. New material combinations introduced unprecedented problems. Leaks progressively attacked concealed elements supporting the stone cladding. The resulting internal problems damaged the stone, damage often misdiagnosed as weakness of the stone material.

---

[1] Principal, THP Limited, *Building Envelope Design and Rehabilitation*, 100 E. Eighth Street, Cincinnati, OH 45202 and Adjunct Associate Professor of Architecture, College of Design, Architecture, Art and Planning, University of Cincinnati.
[2] Principal, The McDonald Group, *Consulting on Stone Masonry Rehabilitation*, 404 Stone City Bank, Bedford, IN 47421.

# Introduction

In engineering's evolution from an empirical process to a rational one, the concept and design of stone facades has changed dramatically. Contemporary architectural designs utilize materials and methods vastly different from masonry's origins, though their character is often intended to look the same. This development of stone usage has not proceeded without pitfalls, especially since we expect new buildings to correct past problems, to perform better, and to last longer. However the opposite is occurring, primarily due to designers delegating responsibility to others.

Wall construction for new buildings tends to be veneered (or "clad") with increasingly thin stones in an effort to reduce cost of cladding and its support. The effect of loading and weathering on such walls is less benign than on thicker, conventional masonry walls *not just* because the stone is thinner, but because the wall systems react to these forces in ways not foreseen. Conventional stone clad masonry walls stack thick stones on ledge angles at each floor, bracing the stones against backup brickwork behind (Fig. 1). Modern stone clad walls anchor stones to metal frames or concrete panels connected to the building's main structure. The thin stone panels span rather than stack (Fig. 2). To *assure the durability of stone facades in new construction*, we must properly apply modern cladding technology to predict and resist these forces.

Many designers do not seem to understand the importance of weathering forces and construction methods, but want the credit for beautiful facades. They are then too quick to blame others when they leak or break. They need to understand that the success of any stone clad facade is dependent upon the design of the attachment details. It is possible to reassign contractor responsibility for detail design back to the designers as part of a rational design program to improve quality, and thus trade liability for durability.

Exacerbating the problem of inconsistent technical leadership from designers, too often, designers, building owners, contractors, and tradespeople proceed along mutually exclusive paths on the same project. As a result, many constructions are of inferior quality. Unfortunately and unnecessarily, that poor quality costs more to build and maintain, providing short service lives. The consequent poor performance and low value of many contemporary facades can be avoided by tightening document content. Not to be confused with demanding more stringent tolerances or higher material strengths, tightening content means concisely communicating structural and visual intent *and detailing how it is achieved.* It means the inclusion of relevant practical experience by describing construction task sequences and procedures. A truly rational process for detailing intent to produce durable designs involves a number of pre-planned, thoughtful, and reasonable steps. Those steps:

- *Determine functional performance of the facade* (the construction's expected useful life, integrity of weather barrier, user comfort);
- *Design buildable architecture* (match expectations to good exemplars, establish aesthetic qualities, study fabrication and installation options to maximize economy and simplify support);
- *Evaluate exemplar buildings* (those which use similar stones, similar support systems, exist in similar environments, and whose performance can be charted);
- *Use standards and references that apply* (such as those developed by American Society for Testing and Materials (ASTM) and recognized technical trade associations which outline design procedures, studies, systems, materials, and conditions related to the project).

In this process, a number of seemingly conflicting interests must be negotiated, meshed, and resolved. The needs of the building owner, the imagination of the architect, and the capabilities of fabricator and builder may not be compatible with the project's schedule and budget without an expert to resolve the conflicts. The engineer may establish criteria at odds with the builder's planned procedures. Issues of motion-tolerant connections must be melded with stone size, shape, handling, support, and anchorage limitations. Tolerance variations between differing materials must be accommodated. Stone's natural strength and appearance variations must be taken into account. Moisture's inevitable entrance into the construction must be recognized and dealt with.

Steps in the program confront barriers between trades and professions, but they can be overcome when the original design is developed in a comprehensive, thoughtful process. Far from compromising architectural creativity, the process enhances it. It improves facade quality and decreases first cost by eliminating redundancies, adding efficient practices, and avoiding problems between trades. It also minimizes risk of diluting design intent and cheating potential durability by keeping structural and visual control of the design with the designers (Fig. 3). Compare this to designers' usual approach

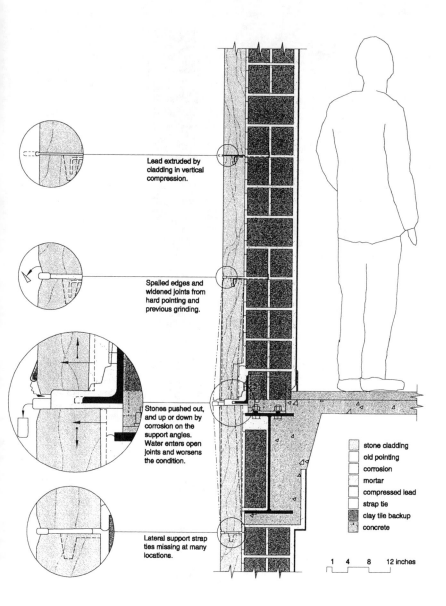

Lead extruded by cladding in vertical compression.

Spalled edges and widened joints from hard pointing and previous grinding.

Stones pushed out, and up or down by corrosion on the support angles. Water enters open joints and worsens the condition.

Lateral support strap ties missing at many locations.

stone cladding
old pointing
corrosion
mortar
compressed lead
strap tie
clay tile backup
concrete

1    4    8    12 inches

**Figure 1: Conventional Stone Cladding**
Non-loadbearing thick stone panels stacked onto angle at floor, braced by backup wall.

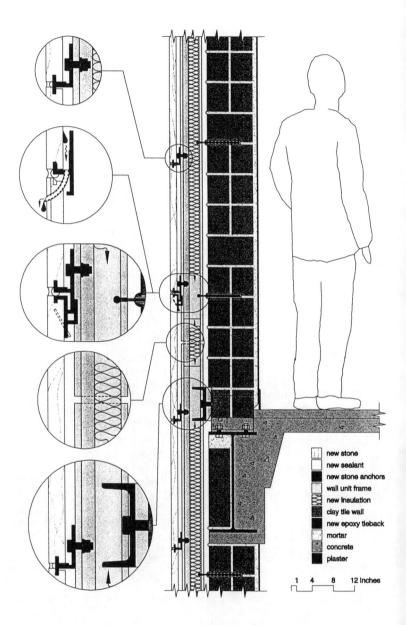

**Figure 2: Modern Stone Cladding**
Thin stone panels are independently anchored to unitized framing hung from each floor.

which fails to show anchoring systems on their documents so bidders can use their own method.   Omitting support detail allows for varied interpretations and visual changes that limit performance and economy because the contractor's method is not coordinated with other systems nor is it coherent with design intent.   Costs are arbitrarily added to cover unknowns.   That method is either arbitrarily *value-engineered* or *un-engineered* to give the contractor a competitive edge, shackling the designer to the bid price with whatever it's thought to buy.

### System Selection (Conceptualization)

**1. Utilize Specialists**
(add stone facade expertise during concept development)

**2. Design Buildable Architecture**
(establish realistic criteria for facade performance and quality)

**3. Evaluate Exemplar Buildings**
(investigate existing buildings to study their features and assess their durability)

**4. Apply Appropriate Standards**
(research, cite, and include pertinent published industry practice)

### Design and Engineering

**5. Generate Concept**
(compile exemplars and standards research with project's architecture)

**6. Refine and Value-Engineer System**
(compare different cladding layout and support options for cost and performance)

**7. Show Fit**
(draw all parts by all trades and describe how they assemble on the Contract Documents)

### Construction (Realization)

**8. Marry Trades**
(submittals from separate contractors must show fit to others' contiguous work)

**9. Test Prototype Mock-Ups**
(verify buildability, prove conformance, establish workmanship standards, finalize engineering)

**10. Educate Tradespeople**
(convey design intent and critical issues, particularly where different trades interact)

**11. Inspect Installation**
(observe construction so project work stays consistent with tests and submittals)

**Figure 3: Process for Assuring Durable Stone Facades in New Construction**
Effective facade system *selection* (conceptualization) through construction (realization) controls intent to achieve quality, doesn't delegate responsibility or sacrifice durability.

## Utilize Specialists...Early!

The rational process suggested herein involves the skills of persons who may not be on the designer's payroll, and knowledge which may not be in his library. *Utilize specialists* to select materials, design system concepts, and detail parts. Expanding the knowledge and information horizons of the design staff and the client along the suggested lines will benefit all concerned parties with a constructive, positive relationship that produces durable, attractive buildings with long and useful lives.

Only facade engineers who have dissected the internal cancers of failed terra cotta, stone, and brick walls understand and respect the destructive potential of water and oxygen on building components. Heat, cold, wind, pollution, salts, vapor, wind, gravity, and inventive architectural forms also contribute to early failure. Sufficient experience with problem-prone shapes, poorly-fitted parts, corrosion-susceptible metals, missing weather barriers, and hard-to-trace leaks will ultimately confer wisdom. Intelligent designers consult with people having this wisdom before they become inseparable from their ideas by selling romantic visions to the client and then by attempting to draw imaginary details for them in contract documents. Remember, stone's modern heritage is very young, and is poised to emerge from empiricism to a rational engineering process.

If the substantial thicknesses and simple, carefully designed and assembled stone systems of the 1920s and 1930s develop serious problems after only sixty or seventy years (Figs. 1, 4 and 5), what can we expect from wall-paper thin, poorly-designed, complex and carelessly assembled stone facades? Failure before occupancy. Though hidden and trapped between interior and exterior finishes, leakage and unrestrained movement progressively weaken the wall. Making thinner walls better requires learning from past construction (Figs. 2, 4, 5, 6 and 7).

A century of explosive urban construction and premature degradation have led us to expect building envelopes to fail too soon. Elsewhere, ancient stone edifices have endured for thousands of years, yet American structures a fraction their age are termed *historic* because few are likely to last that long. Extending durability means changing expectations and changing how facades are designed. A century ago architecture, structure, and weather barrier were provided by massive blocks stacked into thick load bearing walls - one system built by one trade under one master builder (Figs. 6 and 7). Now, stone is most commonly used as a thin cladding. It is non-load bearing and separate from structure and weather barrier, usually built within an assembly having many parts put together by unrelated trades under their own uncoordinated leaders who are disinterested in the *whole*. Meanwhile, designers ultimately responsible for the completed *whole* naively trust this incoherent process and habitually repeat it, believing it protects their professional liability.

**Figure 4: Ohio Departments Building**
Marble-clad brick on steel frame finished in 1932, severely deteriorated due to corroding support.

**Figure 5: Stone Broken Loose**
Corroded supports opened joints, squeezed stones, brook them loose.

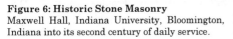

**Figure 6: Historic Stone Masonry**
Maxwell Hall, Indiana University, Bloomington, Indiana into its second century of daily service.

**Figure 7: Solid Bearing Walls**
Blocks of limestone, hand-cut and stacked on site, support building.

Modern building facades are thin walls assembled of varied specialty systems performing independent functions (Fig. 2). A steel or concrete frame supports the building. Cladding on secondary support lined with insulation, waterproofing, and inside finish covers the frame. For this mixed material, multiple-component construction to be durable, parts with different behavior and tolerances must be assembled to allow for their differences without sacrificing performance of other parts. The goal is to mix systems that help each other to perform *more than their primary function*. Few walls have efficient primary systems let alone components that perform several or share functions, because those components are typically conceived to satisfy a specific subcontractor's performance criteria. An experienced facade consultant specializing in stone makes a high-performance stone-clad facade as simple as traditional masonry by bridging the trades and combining functions *during system selection* (Fig. 3).

## Design Buildable Architecture

All parts of a system must perform well with adjacent systems to the same criteria. Where adjacent systems are built by separate trades, these connections must be coordinated to assure all parts conform. The overall rules governing the intended quality of the facade are termed *performance criteria*. They apply after assembly and for the life of the building. Designers escape detailing stone facades by specifying written criteria the construction must meet; many are ignorant of why the criteria exist or how to verify conformance. Though ridiculous, disguising omissions of details with performance criteria is believed to divert blame for failures from designer to constructor.

Any successful design performs and endures by *matching architectural intent to proven construction*. It is how lasting structures have been built since the pyramids. Minimum structural performance, meaning resistance to wind, seismic, rain, snow, deformations and movements will be defined by code and typically measured by analysis. Weather barrier performance, meaning resistance to climatic forces affecting occupant comfort will be defined by building use, orientation, and geographic location. Building value will initially be defined by its cost and purpose, a value that will depreciate if the building is *perceived* to weather ungracefully. Surface ugliness prompting that perception may be a sign of internal weakness that progressively becomes unsafe. In the worst situation, ugliness doesn't occur until concealed conditions are dangerous.

The wise designer will not compromise long-term comfort and looks by omitting details showing how to achieve them. He will describe all aesthetic features, study combined fabrication and installation options to maximize economy, and engineer support that creates the system. The cladding specialist can help designers tailor coherent criteria for related trades so clean, consistent contractual boundaries can be executed (Figs. 8 and 9). Concise details clearly showing contractors' duties need to foresee the sequence of the work and keep it simple and consistent through the project. An installer with two weeks to bid it will comprehend the expected quality of work by how it is presented.

**Figure 8: 190 South LaSalle, Chicago**
Complicated granite-clad gable atop 42-story tower, 1987, thin panels on framing.

**Figure 9: Adjustable Rake Frames**
Support framing formed surface profiles by infilling between structure and stone.

## Evaluate Exemplar Buildings

Durability is the resistance to harmful weathering over time. Though in part subjective, the best measure of a facade's *durability* is its performance on existing buildings. Individual materials for a new facade cannot be evaluated without considering the influence of other system parts. No laboratory can properly assume the conditions and then duplicate the effects of the real world on the system in *real time*. The physics and mechanics of weathering by the combination of changing environmental effects and unknown future conditions on complete building walls prevent tests alone from predicting durability by using accelerated laboratory procedures. Expert analysis of cladding systems which use similar stones, similar support systems, exist in similar environments, and whose performance can be charted, is critical to objectively evaluating a design before it is built.

Although earlier eras of architectural philosophy and design had their problems, the serious and early facade failures we see in today's designs were not typically among them.    Thousands of bearing-wall buildings from antiquity are still extant, and some are still in daily use use in other countries.   Examples of bearing-wall construction from the mid- to late 19th Century are still in use in the U. S., and many are much admired (Figs. 6 and 10).

We learn something about human scale in design from many early buildings, although this article is not intended to address issues of aesthetics.    And coincidentally, the multi-surface exposure of statuary and other architectural ornament often presents text-book examples of a material's durability--if fine detail remains in early examples, a designer using that material today in similar exposures and mass can expect similar if not identical weather-resistance in a contemporary facade.   The designer may have to worry about joints and complex connections, but at least one huge concern may be answered by this reasonable use of exemplar buildings and materials (Figs. 10, 11, 12 and 13).

**Figure 10: Monroe County Courthouse**
Built of Indiana Limestone in 1907, statue grouping at south entry on projected lintel.

**Figure 11: Carved Detail Still Crisp**
Lower element in group shows only minor surface fretting in natural water flow.

**Figure 12: 1820 Limestone Seal**
This original Indiana University seal is nearly unchanged after almost 200 years.

**Figure 13: 1932 Marble Carving**
Detail is clear and sharp after 70 years of urban exposure, 20 since sandblasting.

We also learn something about construction methods from many early buildings, although many of the best examples of improper work are recent projects trying to look like old ones (Figs. 14 and 15). Correcting the poor-performing example's problems results in a well-performing condition (Figs. 16 and 17).

- Use exemplars to select the cladding system for the new project. Use the same or similar stone materials, anchors, and systems that have performed well. Modify them to meet unique project conditions to correct problems in bad exemplars while preserving the features that made them perform well. Insure component compatibility by matching cladding, backup, and support that worked well. Insure good workmanship by employing simple installation methods and skills familiar to local labor. Have the cladding specialist diagnose performance in these areas:

- *Durability* (Assess condition of overall system, then cladding, anchors, support, weather barrier, adjacent envelope elements, maintenance history, and effectiveness of maintenance);

- *Appearance* (Examine stone surface arrangement, orientation to precipitation, joint sizes and locations, panel sizes and shapes, finish, veining, contrasts, color, range, and changes from original);

- *Support* (Check relationship stone to anchors to intermediate support to primary building structure, adjustibility, loads, stiffness, corrosion, signs of problems, fit with weather barrier);

- *Anchorages* (Closely inspect stone anchor engagement into stone; size, frequency and shape; adjustibility, planned movement, permanent set, connection to support, physical condition, interaction with support and tool access);

- *Weather Barrier* (Inspect air, water, and thermal barrier continuity, materials, apparent ease of installation, present integrity compared to original construction, penetrations, support, secondary drainage system, effective comfort of occupants);

- *Installation* (Determine means and methods used to construct the facade, if possible; degree of standardization, unitization, handset, access, and evidence of field modifications).

**Figure 14: Poor Performing Condition**
Miters, return legs, and bullnoses showed problems, thus an exemplar to be avoided.

**Figure 15: Epoxied Profiles**
Glued-and-pinned shapes may save cost at fabrication while reducing durability.

**Figure 16: Well-Performing Condition**
Offsets, slopes, and steps framed show no problems, thus an exemplar to be followed.

**Figure 17: Assembled Profiles**
Separately-set stones on profiled frames are more durable, but cost more to erect.

# The Value of Exemplar Buildings Demonstrated

The value of exemplars is well-illustrated in a study[3] made for an entirely different purpose.  A leading trade association, the Indiana Limestone Institute (ILI), Bedford, Indiana, represents the interests of quarriers and fabricators of Indiana Limestone.  Though Indiana Limestone has endured for nearly two centuries on monumental structures throughout this continent, their concern for acid rain and its putative effect on building stones, particularly those with high lime content, led the organization to study limestone and marble buildings at the epicenter of acid rain damage--cities in southern Vermont.

Recognizing that their interest in the study's outcome might seem to color its results, ILI requested and received unsponsored help of a state government organization, the Indiana Geological Survey (IGS), in both setting up the study, and in analyzing its results.  After visually inspecting buildings constructed of calcium-based stones and obtaining specimens of weathered and protected material, they tested and compared the stones' capacities for crushing and bending.  The study revealed, *contrary to common wisdom and expectation,* the limestone (Fig. 12) and marble (Fig. 13) *suffered little* if any damage as a result of acid precipitation.  Secondly, certain lime-based stones actually *increased in strength* from weathering.

Similarly, investigation of extensively spalled marble cladding on a central Ohio building, initially thought to be material deficiency, revealed dilapidated support and not enough room for movement to be the cause of damage (Figs. 1, 4, 5, 20, 21, 22 and 23).  Here, assessing a poorly-performing exemplar proves valuable in affirming the stone material, but not its support.

These examples show the value of using exemplar buildings instead of relying on scientific speculation or simply observation: The ILI and IGS study high-light successful, durable buildings, utilizing materials which showed a distinct ability to resist the worst possible atmospheres for materials of their class.  Investigation of the damaged exemplar concluded the stone was fine, but the support was bad.

---

[3] Conclusions of study are published *as ILI Technote on Durability and Weathering in Contemporary Atmospheres* available from the Indiana Limestone Institute of America, Inc., ph (812) 275-4426.

## Apply Appropriate Standards

Strategically combining conclusions from specific exemplar studies with general industry standards is the key to an engineering program that will produce a durable cladding system. Using recognized quality standards suggests the need for expertise from industry specialists in the project. Be aware that because they are general in nature, not all standards are pertinent to every project, so cite only those that apply. Ask your specialist to list standards that establish performance and methods without adding unnecessary time and cost to the project. Following are citations from ASTM, the international body of experts promulgating standards for materials, systems, and practice to establish *minimum* quality standards. ASTM's Committee C 18 on Dimension Stone, comprised of stone producers, installers, architects, engineers, scientists, and specialist consultants, generates consensus standards for the safe use of stone on buildings.

In this program, stone cladding material and support systems must meet specific minimum characteristics listed in ASTM specifications and match the performance of exemplars. Standard test methods using small samples establish conformance to minimum strengths. Require only tests measuring properties and variability critical to the use. New tests probably are not needed when recent performance history comfortably proves the stone's capacity. Tests commonly important to modern uses are:

- C 97      *absorption*, sometimes relates to system durability;
- C 99      *modulus of rupture*, stone shear/bending strength;
- C 170     *compressive strength*, if panel isn't likely to buckle;
- C 880     *flexural strength*, strength of panel in bending;
- C 1354    *anchorages*, capacity of stone-to-support attachment;
- C 1201    *chamber test*, full-size panel with anchorages (Fig. 18);
- E 330, 1  *prototype mockup test*, system performance (Fig. 19).

It is true that quarry content and characteristics change as production removes portions of the deposit. For instance, the characteristics and quality of Vermont Marble from 1903 production may not be dependably present in Vermont Marble from 1997. Comparing early strength statements from producers (or specimens from early buildings tested for compression and bending), with specimens from current quarry production, will very quickly establish whether the designer is justified in putting faith in the performance of the older building as an exemplar for the new one.

Meeting minimum strengths does not guarantee safety, for good performance requires system integrity to be confirmed after the parts are assembled. Studying exemplars can show which systems are favorable selections. Design, engineering, and installation of the selected system should verify the structural capacity and constructibility of the design. Guides such as ASTM C 1242 *Standard Guide for ... Stone Systems*, ASTM Manual 21 *Modern Stone Cladding*, the Marble Institute of America's *Design Manual IV*, and the *Indiana Limestone Handbook* explain how to refine and engineer stone facade designs.

**Figure 18: Panel Chamber Test**
When project circumstances require, panel-anchor interaction is measured here.

**Figure 19: Full-Wall Mockup Test**
Build integrated stone and glazing walls full size to verify structural compatibility.

## The Value of the Process Demonstrated

Extending the life of a deteriorating building is one of the most valuable services an engineer can provide. Rehabilitating Columbus' Ohio Departments Building's marble facade illustrates how the process resurrected the historic landmark. The process shrewdly integrated historic preservation concerns with progressive curtainwall philosophy to revive the government seat's original architectural majesty. So persuasive were the benefits of quicker construction start, cost savings, and raised building performance, that the State altered its business-as-usual administration to facilitate the process. Based upon the exemplars studied for the design, these benefits should last many generations.

Severe damage by weathering made the cladding unstable, unsafe for passersby on adjacent sidewalks. To make the building safe, the facade's structural integrity had to be restored. To feasibly extend the building's service life to justify structural rehabilitation, its weather barrier had to be upgraded and its original appearance replicated economically. Creative blending of exemplars, not savvy politics or preservation pressure, showed the upgraded facade made saving the building economically desirable. Five years ago doomed for demolition, its resurrection will make it the most desired address in the state's government.

Completed in 1932 as the State's primary administrative hub, sixty-five years of neglect and bad repairs degraded the historic building. By last year, much of its 150,000 square feet white marble facade was crumbling and dangerously unstable (Figs. 20, 21, 22, 23, 24 and 25). Excavations found failing concealed supports, not poor stone material as first alleged. Development of various repair and replacement options concluded a combined program would permanently correct the problems *plus* improve performance to modern standards... *less expensively than conventional short-term repairs.*

**Figure 20: Facade Investigation**
Close inspection using swingstages showed the severity and quantity of the distress.

**Figure 21: Sample Facade Area**
Note edge spalls, open mortar joints, broken, shifted stones, leaked rust.

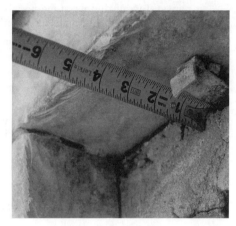

**Figure 22: Sample Excavation**
Several slices of cladding were removed to expose and assess concealed construction.

**Figure 23: Strap To Backup Wall**
Steel straps embedded in the tile backup wall engaged some top gouges in stone.

**Figure 24: Corroded Support Angle**
Leakage corroded the angles, compressed the courses and pushed the stone outward.

**Figure 25: Missing Lateral Tie**
Open gouges showed entire stone courses had no lateral connection to backup wall.

Totally upgrading the building's envelope made fixing the cladding viable. Restoring its original appearance made it desirable and popular. Combining structural restoration, forensics, stone, historic preservation, and environmental expertise developed these benefits (Fig. 2):

- Restored *safety* and long-term structural integrity by replacing the dilapidated support and stacked courses of stone cladding with a new system of thinner stone panels each anchored individually on metal-framed wall units hung directly from the structure.

- Upgraded *durability* by allowing movement, providing proper load path, controlling leakage, and selectively upgrading construction materials and methods.

- Restored original *appearance* and preserved the irreplaceable carved ornament.

- Improved occupant *comfort* and saved energy by adding a cavity system and insulation to keep interior surfaces dry and temperate.

- Upgraded *seismic* performance by stabilizing the retained masonry backup and finish.

## Durability Engineering Principles

The trend to dependency on testing while ignoring existing construction has produced several generations of failing buildings that are being replaced, often with even worse work, and always at immense expense. Existing stone structures have lasted several millennia, and so could properly designed modern stone cladding systems.

Contrasting interests influence facade design. Architects wish to impress viewers with the building's appearance. Engineers intend the safe construction to last. Owners demand economy without compromising an easily maintainable appearance. To *engineer durability* may seem a Herculean task, when resolution of these conflicts risks satisfying only immediate needs; but in practice, it serves the best interests of all parties, and especially of the construction. Key features good designs and well-performing exemplars share are not dependent upon budget as much as they are dependent upon timely consideration and avoiding re-dos. Here are some simple principles that lead to durable, high-performance, and economical stone-clad facades:

- *Buildings move;* surround every cladding bay with motion-tolerant joints so components don't bind.

- *Stone is heavy;* limit panels to sizes that are easily handleable, fabricatable, and installable to minimize damage.

- *Shed water;* orient joints and surfaces to readily shed precipitation, allowing little chance for infiltration.

- *Plan adjustibility;* allow sufficient tolerance between structure and cladding, and cladding panels for realistic adjustments.

- *Keep system simple;* poor workmanship and high costs vary in exponential proportion to design complexity.

- *Standardize parts;* even if parts are custom, potentials for better accuracy and lower cost improve with more replication.

- *Stone varies;* a natural product, stone's strength and visual characteristics vary; use its entire natural range.

- *Material flaws fracture;* use stone material free of structural imperfections; engineer for its variabilities.

- *Exposure changes components;* temperature, moisture, stress, and differentials of these change components' shapes and volumes; allow movement between parts so they don't bind.

- *Hanging stones fall;* design overhead stone lintels, soffits and ceilings to bear on walls rather than only depending upon hangers.

- *Hairline joints fail;* avoid exposed adhesive-joined hairline joints.

- *Multi-purpose parts reduce system cost;* though separate parts designed to perform multiple functions cost more, they delete material and installation costs of parts they eliminate.

- *Metals corrode;* coat, separate, or isolate metals, especially dissimilar metals, to prevent corrosion and surface growth.

- *Poor materials are undependable;* it costs no more to install quality materials than to install inferior materials; in fact, at times it costs less to install quality materials.

- *Facades leak;* plan evacuation routes for moisture that inevitably penetrates the cladding *before it reaches the interior or sensitive components*; be aware that ultimately, even if due to delinquent maintenance, some joints will fail.

Employing durability principles while selecting and engineering a stone cladding system can provide the most economical and durable facade system possible. A durable stone facade extends the potential service life of a building far longer than any other skin type. Achieving durability requires using features of time-proven exemplars that satisfy durability principles.

The presented engineering program mixes exemplars with modern construction technology fitting for contemporary buildings needing long useful lives. But its proactive approach requires designers to set aside their legal armor and stop delegating facade designs to laborers. An insightful colleague speaking with a half-century of experience in the stone industry most eloquently stated:

> *"A return to the position of clerk of the works would probably put more forensic engineers out of business than any other single step. However, with all professionals busily trying to dodge the legal liability bullet, it is unlikely to occur. A belief that testing strengthens the defense against legal liability will continue to fuel the system which will call for more testing to be followed by more failure."*

Assuring the durability of new stone facades means ending negligent design that gives short service lives. The process shifts designers' technical thinking and ethics. But the results revive the fundamental reasons our profession exists.

## Annotated Bibliography

American Architectural Manufacturers Association, *Aluminum Curtain Wall Design Guide Manual*, Des Plaines, IL, Van Nostrand Reinhold, 1979, 1990.

Brookes, Alan, *Cladding Of Buildings*, New York, Construction Press, 1983.

Brookes, Alan, *The Building Envelope: Applications Of New Technology Cladding*, London, Butterworths Architecture, 1990.

Council on Tall Buildings And Urban Habitat, *Cladding*, New York, McGraw-Hill, 1992.

Donaldson, Barry, ed., *New Stone Technology, Design and Construction For Exterior Wall Systems* (STP 996), Philadelphia, American Society For Testing And Materials, 1988.

Hook, Gail, *Look Out Below! The Amoco Building's Cladding Failure*, from *Progressive Architecture*, Cleveland, Penton Publishing, February, 1994.

Indiana Limestone Institute of America, *Indiana Limestone Handbook, 19th Ed.*, Indiana Limestone Institute of America, Bedford, IN, 1992.

Larson, Gerald R., *The Iron Skeleton Frame:Interactions Between Europe And The United States*, in Chicago Architecture, 19872-1922, Birth Of A Metropolis, J. Zukowsky, ed., Munich, Prestel-Verlag with The Art Institute of Chicago, 1987.

Lewis, Michael D., *Anchorage Design: An Installer's Perspective*, from *Dimensional Stone*, Woodland Hills, CA, for Dimensional Stone Institute, Inc., April and June 1992 Issues.

Lewis, Michael D., *Choosing Stone Cladding for Building Facades*, Troy, MI, for Stone World Magazine, March 1993 Issue.

Lewis, Michael D., *Logical Design and Engineering of Stone Building Facades*, Troy, MI, for Stone World Magazine, April 1993 Issue.

Lewis, Michael D., *Modern Stone Cladding, Design and Installation of Dimension Stone Systems*, (Manual 21), Philadelphia, PA, American Society of Testing and Materials, 1995.

Lewis, Michael D., *Planning & Designing Stone Cladding*, Proceedings of CIB-ASTM-ISO-RILEM International Symposium on the Performance Concept in Buildings, Haifa, Israel, National Building Research Institute, 1996.

Nowak, Andrzej and Galambos, Ted, ed., *Making Buildings Safer For People*, New York, Van Nostrand Reinhold, 1990.

McDonald, William H., *A Short History of Indiana Limestone*; Bedford, IN, Lawrence County Tourism Commission, 1995.

O'Connor, Jerome P. and Peter Kolf, *High-Rise Facade Evaluation And Rehabilitation*, from Concrete International; Detroit, MI, American Concrete Institute, September, 1993.

Shadmon, Asher, *Stone, An Introduction*, London, Intermediate Technology Publications Ltd., 1989.

Tuchman, Janice L., *Curtain Walls In The Spotlight*, from Engineering New Record, New York, McGraw-Hill, December 13, 1993.

Wilson, M. and Harrison, P., *Appraisal And Repair Of Claddings And Fixings*, London, Thomas Telford, 1993.

Winkler, Erhard M., *Stone: Properties, Durability In Man's Environment, 3rd Ed.*, Munich, Springer-Varlaq, 1994.

# SUBJECT INDEX

Page number refers to the first page of paper

# AUTHOR INDEX

Page number refers to the first page of paper